This book belongs to

____

Who wants to become

A/An ______________________________

By (enter date) ______________________________

Student's signature ______________________________

Date signed ______________________________

Parent's signature ______________________________

Date signed ______________________________

10 9 8 7 6 5 4 3 2

First published 2022 with a non-exclusive licence from the author to CHEETAH® Purrrrrrr Publishing ('CHEETAH®'), an imprint of CHEETAH® Toys & More, LLC.

ISBN-13: 978-1-7328369-5-2

ISBN-10: 1-7328369-5-2

Permission request(s) should be submitted to the publisher in writing at one of the addresses below:

Contact information:

CHEETAH® Toys More, LLC.
207 Main Street, 4th Floor
Hartford CT 06016

Port Antonio P.O.
Portland, Jamaica

info@mycheetahacademy.com
paulettetrowers@yahoo.com
876-909-6311 (WHATSAPP ONLY)

Authors: Paulette Trowers, Juris Doctor and a team of educators
Editors: Fiona Porter-Lawson, Patricia Bryan, Alisa Grant
Cover and interior design: CHEETAH®
Publisher: CHEETAH®

ACKNOWLEDGEMENT

Thanks to my Creator for giving me all that I needed (persons with the expertise, coupled with the finances, time and opportunity) to bring this value-added book, **Primary Exit Profile (PEP) Social Studies Practice Questions Grade 6** workbook, into the hands of the students, teachers and parents/guardians.

Thanks also to all persons who helped to make this book a reality. This publication took longer than expected, but 'a dream deferred is not a dream denied.' It took a dedicated team of authors, editors, illustrators, book designers/ a book designer, Ministry of Education and Youth reviewers, printers, warehouse crew, delivery drivers, photographers, school administrators and parents/ guardians to complete/accomplish the task.

A great deal of knowledge, research, time, imagination, creativity, strong work ethics and collaboration was involved to develop this text from concept to reality and into the hands of the students and teachers.

Special thanks to:

- **Editors:** Fiona Porter-Lawson, Patricia Bryan, Tanisha Dawkins, Alisa Grant, Lucinda Peart
- **Contributors:** Rosemarie Pottinger, Camile Clarke, Delrona McPherson, Marshalee Dale-Powell, Evan Hutchinson, Alisa Grant

CHEETAH® appreciates the Jamaica Information Service (JIS) and the National Library of Jamaica for providing accurate, original and updated content.

The CHEETAH® Purrrrrrrr publishing team is grateful for the opportunity to positively impact education and hopes that this book will enhance the overall teaching and learning experience.

TABLE OF CONTENTS

TERM 2, UNIT 2

TERM 2, UNIT 3

What is Social Studies?
Learn more by watching this video:
https://www.youtube.com/watch?v=e-0NRGMakeY
Come wid mi. Mek wi prep. for PEP and life.

Dear CHEETAH® family ,

Welcome to *CHEETAH® PEP Social Studies Practice Questions* workbook. This book is designed to prepare you for the PEP (Primary Exit Profile) exam and for life in general. You will notice that this book is arranged in the following order:

- A summary of the key concepts according to the learning objectives that you are expected to master
- Practice questions (2 questions per learning objective) to be scored at the teacher's discretion
- Full-length tests (3 different tests of 40 questions each)
- Answers and explanations.
- A glossary that defines the key words. These key words are highlighted in colour within the concept section.

Through this book the item types will vary according to the learning objectives. For multiple choice questions, circle the letter for your answer. For some items, instructions such as select, match or choose may be given. In these cases circle the letter beside the correct option.

An Overview of the Topics

Taken directly from the National Standards Curriculum (NSC)

This is what you are expected to learn:

Subject	Term 1	Term 2
Social Studies	**Our Common heritage** The Chinese and East Indians in Jamaica Promoting and preserving Caribbean culture Independence in Jamaica, Haiti and Cuba National heroes - Marcus Garvey, Norman Manley, Alexander Bustamante **Living together** National symbols and emblems	**The physical environment and its impact on human activities** Mountain environments and human activities Landmasses and water bodies of the world Locating places using lines of latitude and longitude **Living together** Decision making at the national level and how decisions affect citizens Rights and responsibilities of citizens

Journey with me, Roy, the Jamaican Rooster. I will educate, entertain and inspire you the CHEETAH way. According to Confucius, 'A journey of a 1,000 miles begins with a single step.'
Yuh ready? Come journey wid mi. Let's prep for PEP and life!

SOCIAL STUDIES

Key Concepts

'Knowledge is power. School is the tower where you get the tools.

To be successful in life, babe, you've got to get to school. Out there in the real world, it's the survival of the fittest, not the prettiest in the show.

How successful you are has a lot to do with how much you know.

You are a cheetah, a future leader. Keep learning more.'

Excerpt from the CHEETAH® Theme Song

SOCIAL STUDIES KEY CONCEPTS

Welcome to CHEETAH® (Connect to Higher Education, Electronic Tools, Application and Help) social studies key concepts. As you learn, I want you to journey with us by taking a few simple steps:

1. Review the key concepts. This is a summary of some important information that you will need to answer the questions in this book.
2. Take the practice questions.
3. Evaluate your work and have someone evaluate it for you.
4. Take the full-length 40-questions tests.
5. Have fun while you learn about Jamaica, our neighbouring islands and the world around us.

Don't forget to take your imagination and work ethics with you.

Can you imagine what your knowledge and test scores can be if you are prepared?

Remember the six Ps:

Proper

Planning and Preparation

Prevent

Poor

Performance.

Unknown.

Come wid mi. Let's prep for PEP and life!

TERM 1, UNIT 1

Promoting and Preserving our Caribbean Culture

As you review the information in this unit, think about these items:

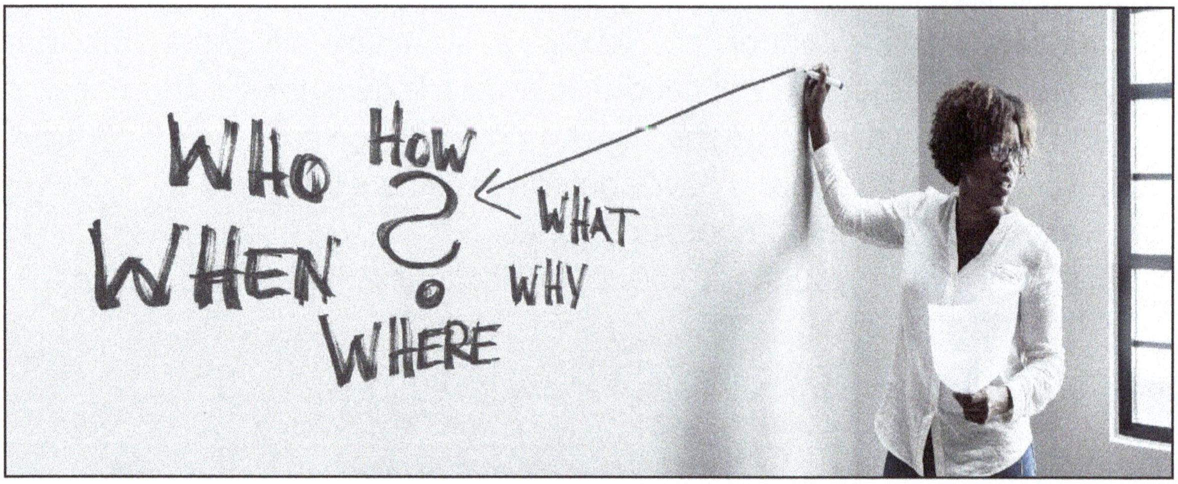

Who came to Jamaica?

Where did they come from?

When did they come?

Why did they come?

What was life like for these different groups on the plantations?

How have these groups influenced our current way of life?

TARGET

Key concepts to know and apply

- **indentured servant**
- **indentureship**
- **contract**
- **festival (carnival)**
- **immigrant**
- **migration**
- **push and pull factors**
- **culture**
- **heritage**

You will find the definitions in the glossary.

The Early Immigrants to Jamaica

Indentured servants from places such as India, Germany and China were **contracted** to work in Jamaica after slavery ended. The period of **indentureship** lasted up to five years. Many of these **immigrants** — persons who move from one place to settle in another — left their homes because of different **push factors** such as poverty, drought, unemployment or persecution.

They were seeking a better way of life. They were offered jobs, freedom and fertile lands if they migrated to Jamaica. These incentives are called **pull factors**.

Culture, Heritage and Ethnic group

When the **immigrants** of different ethnic groups came from their homelands to settle in Jamaica, they brought with them many cultural practices, customs, traditions and items of great significance to their new community. They held **festivals (carnivals)** to celebrate different cultural dates and activities.

- **Culture** describes the way of life of a people that includes their customs, traditions, festivals and beliefs.
- **Heritage** refers to knowledge, cultural practices and items passed down from generation to generation.
- **Ethnic group** is a group of people with a similar physical appearance, language, religious practices, ancestors and culture.

These festivals were religious (Diwali, Hosay, Easter and Christmas) or secular (Carnival Chinese New Year and Crop over).

Figure 1: *Chinese and East Indian immigrants came to Jamaica. Inspired by https://nlj.gov.jm/history-notes-jamaica.*

TARGET

The Arrival of Different Ethnic Groups to Jamaica

Ethnic groups of the Caribbean mainly originated from Africa, Europe and Asia. These ethnic groups include the Europeans, Africans, Indians, Jews, Germans and Chinese.

Ethnic Group	Date of Arrival in the Caribbean	Date of Arrival in Jamaica
Spanish	1492	1494
Jews	(unknown)	1494
Africans	1504	1513
British/English	1623	1655
Chinese	1806	1854
Indians	1834	1845
Germans	(unknown)	1834

Push and Pull Factors for Indentured Servants

Different groups came because of unpleasant situations at home, such as poverty, unemployment, drought and persecution. These situations are called **push factors**. Other immigrant groups came here because of promises of access to housing, food, fertile land, medical care and freedom. These are called **pull factors**. These groups were able to set up commercial businesses, such as shops and supermarkets, at the end of their indentureship.

Plantation Life for East Indian and Chinese Immigrants

The **indentured servants** were encouraged to migrate to Jamaica with the promise of good wages, houses, medical care, food and clothes. Life in Jamaica was not what they expected, as they were forced to work long hours and were paid meagre wages. They hardly received medical care, so many got sick and died from diseases. Their homes were small, hot and had dirt floors. They were not allowed to leave the plantation, even though they had been promised freedom.

Relationship Among the Ethnic Groups

Cultural differences affected communication among the groups, and this often led to conflicts. The freed African saw the Chinese as a threat, and the Africans treated the East Indians as inferior because the East Indians received less pay than the ex- slaves did. The East Indians saw themselves as superior to the Africans because their caste system valued lighter skin.

The relationships among the Europeans and the Africans and their descendants did not improve after slavery. The Europeans considered themselves superior to the Africans, even if the Africans were wealthy and educated. Europeans continued to treat Africans and their descendants brutally.

As the years passed, the relationship among these groups improved significantly. Different ethnic groups participated in cultural celebrations and festivals and enjoyed one another's food and music. Some were converted to new religions, especially to Christianity. There was the beginning of the practice of syncretism involving the mixing of religions. Interracial marriages also became more common.

Figure 2: *Immigrants from different ethnic groups who came to Jamaica.*

Economic Contributions of the East Indian and Chinese Immigrants

The East Indian and Chinese have made significant contributions to the Jamaican economy, creating businesses and employment for many Jamaicans, as shown in the table.

Contribution of East Indians and Chinese to the Jamaican Economy	
East Indians	**Chinese**
farming, medicine, horse racing, manufacturing, wholesale and retail businesses, politics	wholesale and retail businesses, such as grocery stores, laundromats, bakeries, restaurants

TARGET

Immigrants' Cultural Goods and Services

Figure 3: *A sample picture depicting Diwali, the Indian festival of lights.*

Culture is the way of life of a people and includes language, dress, food, festivals, music and social and economic activities. Many of our cultural goods and services help the economic development of the country. Our rum, music and coffee are examples of cultural goods that promote tourism and contribute to the economy.

Traditions and Celebrations that Influenced our Culture

Ethnic Group or Country	Events	Description
East Indian	Diwali (Divali)	Diwali is the festival of lights. It is a nine-day celebration paying tribute to the goddess of prosperity.
East Indian	Hosay	Hosay is a three-day Muslim festival involving dancing in the streets to the sound of drums.
Jamaica	Easter/Christmas	Christmas and Easter are Christian festivals. Christmas celebrates the birth of Jesus Christ. Easter commemorates Jesus's death, burial and resurrection.
Barbados	Crop Over	Crop Over marks the end of the harvesting of sugar cane. People dress up, participate in parades and visit stalls to support local vendors.
Chinese	Chinese New Year	The Chinese New Year celebration begins in early spring with the arrival of the new moon and lasts for fifteen days. Families get together and eat, dance and enjoy fireworks.

Promoting Caribbean Culture

Caribbean culture is a melting pot; cultures of the islands are mixed, each influencing the other. Preserving and promoting Caribbean culture can strengthen the cultural identity of a people.

Ways to Preserve Culture	Ways to Promote Culture
Have regular ceremonies and celebrations.	Share cultural activities and celebrations as a region.
Take care of cultural landmarks such as buildings, monuments, historic sites.	Teach aspects of Caribbean culture in schools.
Record information and use it to teach the younger generations. Share stories and folk tales with younger generations, write books and create films.	Record cultural practices and activities. Write newspaper articles, make films and documentaries and publish them.
Share our culture with others.	Create more opportunities and facilities for cultural celebrations.
Record experiences and share on public platforms, such as on websites or social media.	

As a region, we can promote our culture by offering more regional activities in sports, music and theatre so visitors can learn about our culture as they enjoy themselves. We can promote Caribbean culture to visitors by offering local foods and entertainment.

The Contribution of the East Indians and Chinese to Caribbean Culture

Although the Chinese and East Indians make up a small percentage of our population, they have contributed greatly to Caribbean culture.

Chinese

The Chinese brought:

- Food products, spices and fruits such as cinnamon and lychee
- Religious practices, such as Buddhism
- Commerce, such as privately owned Chinese grocery shops, supermarkets and restaurants
- Festivities such as the Chinese New Year and the use of fireworks.

East Indians

The East Indians brought:

- Food items, spices and fruits to Jamaica such as roti and dahl, curry and mango
- Religious practices, such as Islam and Hinduism
- festivals like Diwali and Hosay
- Music and dance traditions which are still part of Jamaican culture today.

TARGET

Gathering Information

The Internet has made it possible for anyone to present information to the public. It is important to know how to choose sources that are most likely to give trustworthy information; for example, an organisation or government website will probably provide more credible information than a blog.

Choose sources that have:

- More facts than opinions. If an opinion is presented, it should be well reasoned and based on facts
- An identified author rather than sources that have no named authors
- Evidence that the site is regularly updated, if using a website; references or links to show where they got their information; no evidence of bias; have a balanced view of the subject
- A clear purpose to inform readers rather than to persuade.

TARGET

Economic Value of our Cultural Industries

Our culture not only identifies us as a unique people but also creates products that contribute to the country's economy. Industries such as music, sports, the performing arts, fashion, craft, food and beverage earn national income. Jamaica's economy has improved and has seen benefits such as economic growth, less poverty, more jobs and increased tourist arrivals, resulting in more foreign exchange.

Did you learn anything new about our history?
Michael Crichton said, 'If you don't know your history, then you don't know anything. You are a leaf that doesn't know it is part of a tree.'
Do you agree? We have more units to explore. Come wid mi. Let's prep for PEP and life!

TERM 1, UNIT 2

Our Common Heritage

Key concepts to know and apply

- independence
- colonial rule
- commonwealth
- constitution
- nation
- trade union
- political party
- self-government
- universal adult suffrage
- franchise
- revolution

You will find the definitions in the glossary.

Timeline from the Arrival of the Tainos to Independence

The Tainos settled on the island of Jamaica more than 2,500 years ago. Christopher Columbus saw them here when he arrived in 1494. From 1494 to early 1600, many Spanish settlements were established in Jamaica. The Tainos eventually died from diseases and harsh treatment by the Spaniards. In the 1650s, the Spanish settlements were attacked by the English and Jamaica was captured in 1655.

The English developed a prosperous sugar industry and brought enslaved persons from Africa to work on the plantations. Enslaved Africans were first brought to Jamaica in 1513. As the sugar industry grew, the treatment of the enslaved grew harsher. This led to many revolts and rebellions. Some run-away enslaved, called Maroons, escaped to the mountains and formed settlements.

In 1807, the slave trade between Africa and Jamaica was abolished, and in 1834, emancipation was granted to the enslaved. There was a brief period of apprenticeship to help with the transition to full freedom. Four years later, slavery was fully abolished. Many free Africans left the plantations and settled in different parts of the island. The plantation owners, therefore, turned to

China, Africa, Europe and India for indentured labourers. The first set of Indian labourers came to the island in 1845, followed by the Chinese in 1854.

Workers were unfairly treated, hence, there were many strikes, revolts and riots. This strengthened the desire for **independence**, which started with the Maroons. They fought with the British for freedom in the early 1700s. This struggle for independence continued in 1938 with the labour rebellion and in 1944, Jamaica gained **universal adult suffrage**. Although Jamaicans had the right to vote, the British still held power over the country. Leaders like Marcus Garvey, Norman Manley and Alexander Bustamante promoted nationalism and self-government during the mid-1900s.

TARGET

The Life and Work of Marcus Garvey, Norman Manley and Alexander Bustamante

Norman Manley

Norman Manley was born in 1893. He was a lawyer who fought for the rights of the workers. These rights included better working conditions and more pay. He supported the **trade union** movement that was led by Alexander Bustamante. Manley believed in democracy and universal adult suffrage. He founded the People's National Party, helped to draft the Jamaican constitution and lobbied for independence. He died on September 2, 1969, but was honoured as a National Hero on October 18 of that same year.

Alexander Bustamante

Alexander Bustamante was born in 1884. He was Norman Manley's cousin. He helped with writing the **constitution** of Jamaica and presenting it to the Queen. He advocated on behalf of the people and called for fair wages and workers' rights. He made people aware of the socio-economic problems of the poor in Jamaica. People became frustrated with unemployment and poor working conditions, and this led to widespread riots and strikes. He formed the Bustamante Industrial Trade Union (BITU) to protect the rights of the workers. He formed the Jamaica Labour Party (JLP) in 1943 and became the first Prime Minister of Jamaica in 1962.

Marcus Garvey

Marcus Garvey was born in 1887 and was recognised as Jamaica's first national hero in 1969. He believed in racial equality and black pride. He encouraged **self-governance** and self-reliance among blacks. In 1914, he founded the Universal Negro Improvement Association (UNIA) in Jamaica. Garvey then travelled to the United States but was imprisoned there, then deported in 1916 because of his teachings. He was unsuccessful in his attempt to form a **political party** in Jamaica a few years later. He died in 1940 in England.

Lessons from the Lives of Marcus Garvey, Norman Manley and Alexander Bustamante

Here are some important lessons we can learn from our national heroes.

- We should stand up for our rights and the rights of others.
- We should teach others in a way that empowers them.
- We can achieve our goals by discussion and unity.
- We should look after the interests of others, especially the powerless.
- We should exercise our rights and carry out our responsibilities.

TARGET The Paths to Independence Taken by Jamaica, Haiti and Cuba

Cuba

Cuba was a former Spanish **colony** (formerly governed by Spain). They experienced hardships under this government. Although Cubans staged many rebellions against the government of Spain, it was in 1898, after the Spanish-American war, that Spain withdrew from Cuba. After three and a half years of American military rule, Cuba gained formal independence on May 20, 1902.

Haiti

The Haitian **Revolution** against French **colonial rule** was the highlight of Haiti's independent history. Toussaint L'Ouverture led a slave revolt against the French, but he was arrested and tortured to death. Jean-Jacques Dessalines led another revolt in August 1791, which resulted in the country gaining independence in 1804.

Jamaica's Road to Independence

Jamaica's road to independence started with the Maroons, who fought with the British for freedom. This path continued in the 1800s when slavery was abolished and Jamaica gained adult suffrage. Leaders like Marcus Garvey, Norman Manley and Alexander Bustamante promoted nationalism and self-government.

Workers protested against inequality and fought with the authorities. Political leaders developed a constitution and structure for independence. In the 1950s, the Jamaican government and the British government began the long series of discussions that finally led to full independence in 1962.

The Major Personalities Involved in the Independence Movement in Jamaica, Cuba, Haiti

During the fight for independence in a country, many people play key roles. Sometimes these people are awarded the country's highest honour, such as the Order of National Hero.

Heroes Who Fought for Independence

Jamaica	Cuba	Haiti
• Marcus Garvey • Alexander Bustamante • Norman Manley	• José Martí • Carlos Manuel de Cespedes	• Toussaint L'Ouveture • Jean Jacques Dessalines • Henri Christophe

Commemorating Independence in Jamaica and other Countries

Cuba

Independence Day in Cuba is October 10 each year. However, May 20 is commemorated as the day Cuba got independence from Spain in 1902. It is a public holiday marked by large street parades, music, other festivities and food.

Haiti

Independence Day in Haiti is celebrated on January 1 each year. It is celebrated with fireworks, dancing and renditions of the national anthem, in honour of Jean-Jacques Dessalines. He is hailed as the national hero of the revolt that led to independence. They drink a traditional soup, made from pumpkin, and named Joumou,

Jamaica

Independence in Jamaica is celebrated on August 6 each year. There are dances, parties and street parades with people dressed in the national colours. At the National Stadium there is also a national celebration of cultural presentations, known as the Grand Gala. This showcases aspects of the country's past and the struggles to reach independence.

The Significance of Independence Day

In Jamaica, Cuba, Haiti and other **nations**, independence means freedom from colonial rule, full self-government and complete adult suffrage. It also means the people of the nation can institute their constitution and can ensure equality for all. People can vote freely in a democratic election without interference from another country.

Jamaica's Decision to Pursue Independence

Why did Jamaica seek independence from Britain?

Before slavery ended in 1834, people worked without wages and lived under very difficult conditions. After slavery ended, very little changed in people's standard of living. The colonial government paid little attention to the working class, who had very little say in how they were governed.

Why did Jamaicans need to have a say in government when it was controlled by Britain?

Self-government was important to the people. They would be able to develop policies to improve the economy of the country and people's well-being. The people would be able to choose who they wanted to govern them.

Dependent and Independent Countries in the Caribbean

Independent Nations

Independent Nation	Year of Independence	Gained Independence from
Antigua and Barbuda	1981	UK
The Bahamas	1973	UK
Barbados	1966	UK
Belize	1981	UK
Cuba	1902	USA
Dominica	1978	UK
Dominican Republic	1865	Spain
Grenada	1974	UK
Guyana	1966	UK

Independent Nation	Year of Independence	Gained Independence from
Haiti	1804	France
Jamaica	1962	UK
St. Kitts and Nevis	1983	UK
St. Lucia	1979	UK
St. Vincent and the Grenadines	1979	UK
Trinidad and Tobago	1962	UK
Venezuela	1830	Spain

Dependent countries

Dependent Countries	Governed by
Anguilla	UK
Aruba	Netherlands
British Virgin Islands	UK
Cayman Islands	UK
Curacao	Netherlands
Montserrat	UK
Puerto Rico	USA
St. Barthelemy	France
St. Martin	France
St. Maarten	France
Turks and Caicos Islands	UK
United States Virgin Islands	USA

Showing Appreciation for Independence Heroes

Jamaica's road to independence has been paved by great men of the country. These include Marcus Garvey, Alexander Bustamante and Norman Manley. To show appreciation for their work, Jamaica declared them national heroes, the highest honour of the country. A special day, National Heroes' Day, is celebrated yearly to honour them and others who share the same honour. A park, National Heroes' Park, was established in their memory. Statues of our heroes are prominently displayed throughout the country. Their faces are also printed on our national currency.

Independence Heroes	How We Show Appreciation
Marcus Garvey	-Marcus Garvey Drive is named after him. - Scholarship in his honour for boy with the highest PEP score in a government school - National Heroes Park has his shrine - Face is on our current $100 note and $20 coin - Garvey's statue is in front of the parish library in St Ann's Bay
Alexander Bustamante	- National Heroes Park has his shrine. - Face along with Norman Manley are on the $1,000 note - Face on the $1 coin - Hospital in Kingston, bridge in St. Thomas, and highway in Clarendon named for him. - Statue at the entrance of St. William Grant Park in Kingston

Norman Manley	- National Heroes Park has his shrine. - Face along with Alexander Bustamante is on the $1,000 polymer (plastic type of material) note - Face on the $5 coin - T Norman Manley International Airport and the Norman Manley Law School are named after him. - His birth place, Roxborough Castle Plantation in Manchester, is a National Monument.

TARGET

Jamaica's Decision to Pursue Independence

Why did Jamaica seek independence from Britain?

- Jamaicans wanted to have the freedom to decide on their own laws, foreign trade and defense.
- The British government could not always relate to the challenges of Jamaicans.
- Jamaicans felt that their socio-economic needs were not being met. Wages were low, unemployment high and workers were not always treated fairly. This led to the rise of trade unions and political parties that tried to represent the people's needs.
- There was a rising group of educated, coloured, middle-class individuals who were able to represent the poor and working classes. They educated the people about the needs and benefits of self-government.

Benefits of Remaining a British Colony

- Britain is a wealthier nation and can provide economic support for its dependent territories.
- Small countries like Jamaica cannot defend itself from larger countries in times of war. Being a dependent nation will guarantee protection from the mother nation.

Conflicts in Collaborative Groups

A conflict is a disagreement between/among people. People have different likes, dislikes, opinions and ways of doing things. These differences can cause conflicts and major problems if not resolved.

Here are some questions to think about when facing conflicts while working in groups:

1. What is the disagreement about?
2. What is the goal you want to reach?
3. How can you work together to meet this goal?
4. What are the problems that are preventing you from reaching this goal?
5. What is the solution that you have all come up with?
6. What tasks should each person complete?

Listing Sources

A source provides useful information for doing research. When using information found in material written by others, you should give credit or cite the source. This allows others to know that the ideas are not your original ideas. A list of sources used to complete an assignment or research is called a reference list or bibliography. It includes the name of the author, the tilte of the work (such as a book or an article), the publication date and place or the link to the information if it has been published on the Internet. Internet references also include a date that tells when you got the information from the website.

Here are some examples:

- Bradshaw, C. (2008). Jamaica Then and Now. Portland: JoJo's Publishers Dunkley, A. (1985). *Going back to Our Roots*. New York: Fisher's University Press
- Jackson, N. (2019, February 14). My Journey to Freedom. The Sunday Gleaner. Retrieved June 10, 2020, from https://www.gleaner.com/

There are different styles for formatting sources, but they each have all the information needed to find the referenced source. The list of sources is written in alphabetical order, according to the authors' last names.

TERM 1, UNIT 3

Living Together

TARGET

key concepts to know and apply

- emblem
- flag
- coat of arms
- symbol
- anthem
- crest
- bearing
- motto

You will find the definitions in the glossary.

TARGET

Jamaica's National Emblems and Symbols

In preparation for Jamaica's independence, several national symbols and emblems were agreed on. They would represent the many areas of Jamaica's political, cultural and economic life.

The national symbols depict all the people of Jamaica as one united community. These symbols promote patriotism among Jamaicans. Ideally, they capture the goals, values and history of the nation, and aim to instill pride and unity. They are important to our national identity.

The national **symbols** and **emblems** of Jamaica are:

- The flag
- The coat of arms
- The anthem and national pledge
- The national fruit — ackee
- The national bird — Swallowtail Humming Bird
- The national flower — Lignum Vitae
- The national tree — Blue Mahoe.

The National Emblems

1. The Jamaican Flag

On Independence Day, August 6, 1962, the Jamaican **flag** was raised for the first time. The flag is black, green and gold. The gold diagonal cross runs through the centre of the flag, with the fly and hoist triangles black and the top and bottom triangles green.

Meaning of the Colours of the Flag

The colours of the flag symbolise the idea that 'the sun shineth, the land is green and the people are strong and creative.'

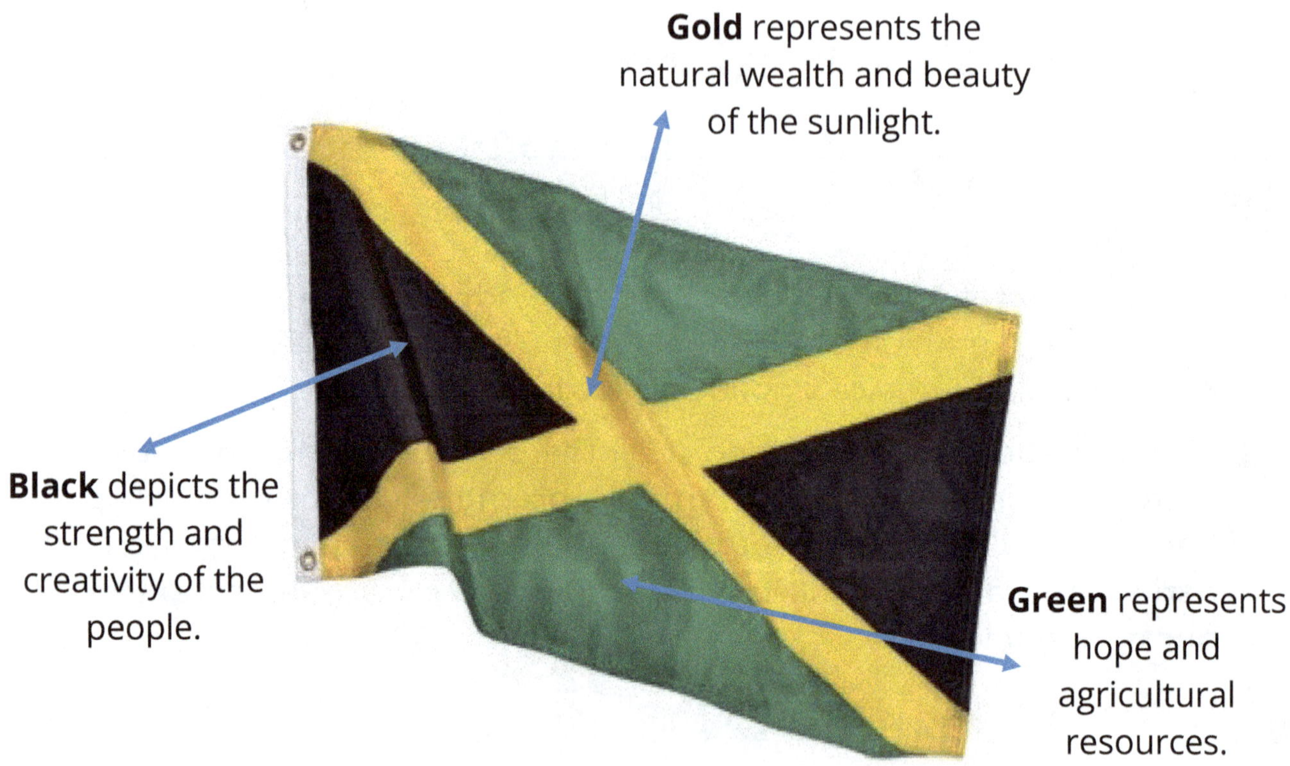
Gold represents the natural wealth and beauty of the sunlight.
Black depicts the strength and creativity of the people.
Green represents hope and agricultural resources.

Source: *https://opm.gov.jm/symbols/national-flag*

Figure 4: *The Jamaican national flag.*

Protocols for Using the lag

Protocols help us know what to do and the rules to follow in certain circumstances.

Here are some protocols to follow when using the Jamaican flag:

- Citizens should respect the flag and treat it as a sacred emblem.
- The flag should never touch the ground.
- Avoid temporarily using the flag as decoration, except on state occasions.
- Burn the flag when it needs replacing.
- The Jamaican flag should never be smaller than any other flag flown at the same time in Jamaica.
- The flag of Jamaica should fly higher and to the right of other flags in Jamaica.
- No flag from another country should be flown in Jamaica unless the Jamaican flag is also flown, except at a foreign embassy, mission or consulate.
- The flag should not be draped over vehicles except on military, police and state occasions.
- As a sign of official mourning, the Prime Minister's Office will declare that the flag is flown at half-mast for a certain period.

Figure 5: *The Jamaican coat of arms.*

2. Coat of Arms

Jamaica was assigned its first **coat of arms** in 1661. Since then, there have been many changes.

The first **motto** on the coat of arms was Latin and read, 'Indus Uterque Serviet Uni'. This was later changed to English, the official language of Jamaica. It is now 'Out of Many, One People'.

Changes to the Coat of Arms

Coat of arms of Jamaica from 1875 to 1906.

Coat of arms of Jamaica from 1906 to April 8,1957.

Coat of arms of Jamaica from April 8, 1957 to July 13, 1962.

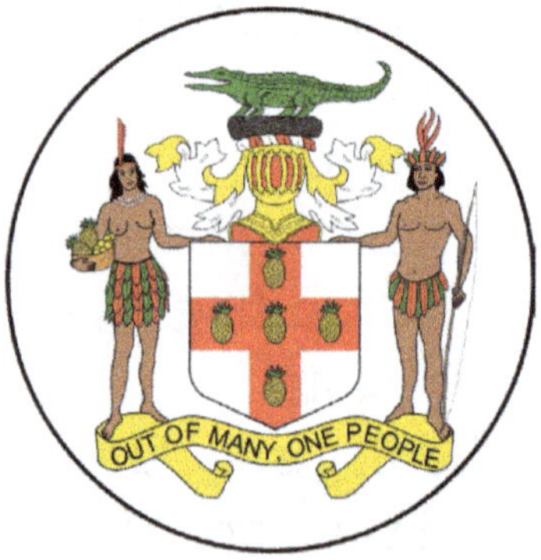

Coat of arms of Jamaica from July 13 to August 6, 1962.

Figure 6: *Changes to the Jamaican coat of arms.*

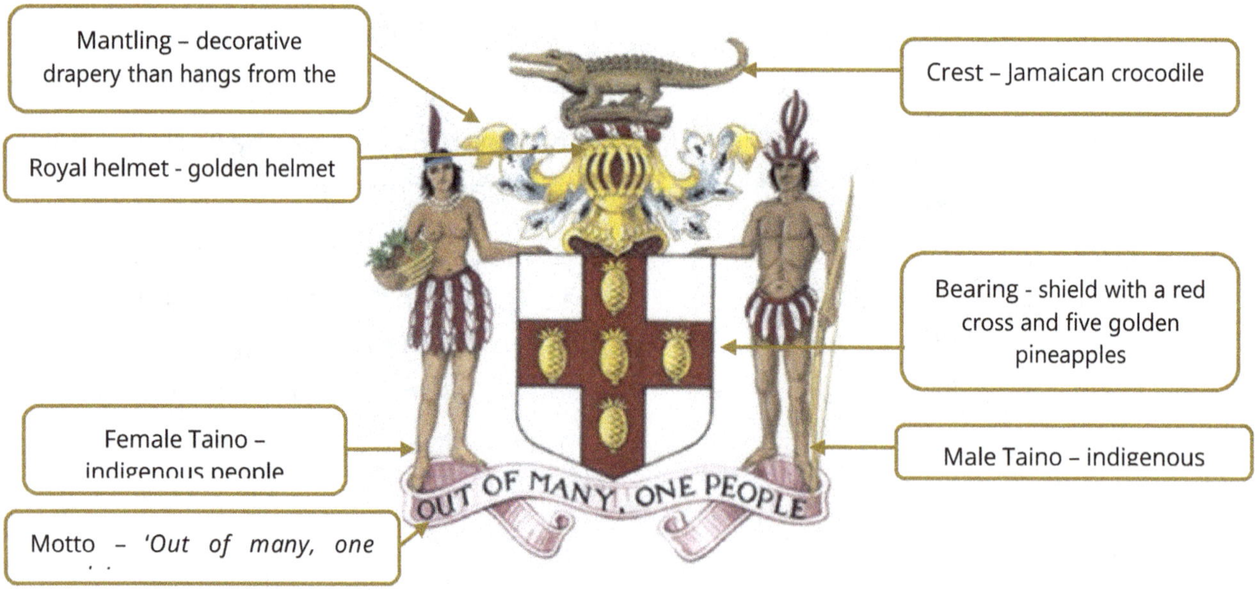

Figure 7 : Labeled diagram showing features of the Jamaican coat of arms.

The coat of arms we now use shows a male and female Taino standing on each side of the shield. The bearing is the central part of the coat of arms to which everything is attached. The bearing on the Jamaican coat of arms consists of the shield with a red cross and five golden pineapples. The **crest** is a Jamaican crocodile on top of the Royal Helmet and mantling. The coat of arms cannot be used without official authorisation from the Prime Minister's Office.

Protocols for Use of the Coat of Arms

- The coat of arms should not be used for commercial or personal reasons.
- It should only be used to identify official documents.
- It should not be imprinted on any article of clothing of any group representing Jamaica locally or internationally.
- It should not be placed on programmes which will eventually be discarded or not treated respectfully.

3. The Jamaican National Anthem and National Pledge

Jamaica's National Anthem

Eternal Father bless our land, Guard us with Thy Mighty
Hand, Keep us free from evil powers,
Be our light through countless hours. To our Leaders, Great
Defender, Grant true wisdom from above.
Justice, Truth be ours forever, Jamaica, Land we love.
Chorus: Jamaica, Jamaica, Jamaica land we love.

Teach us true respect for all,
Stir response to duty's call, strengthen us the weak to cherish,
Give us vision lest we perish.
Knowledge send us Heavenly Father, Grant true wisdom from above.
Justice, Truth be ours forever, Jamaica, land we love.
Chorus: Jamaica, Jamaica, Jamaica, Land we love.

Interpretation of the National Anthem

Jamaica's national **anthem** is a prayer that asks God to bless and guard our country, protect us from evildoers and bless our leaders with divine wisdom. It expresses the citizens' love of Jamaica. The prayer continues to ask God to help the people of the nation to respect each other, be responsible and be compassionate to the weak. Finally, the anthem asks for knowledge and wisdom from God to sustain life.

Jamaica's National Pledge

Before God and all mankind, I pledge the love and loyalty of my heart, the wisdom and courage of my mind, the strength and vigour of my body in the service of my fellow citizens; I promise to stand up for justice, brotherhood and peace, to work diligently and creatively, to think generously and honestly, so that Jamaica may, under God, increase in beauty, fellowship and prosperity, and play her part in advancing the welfare of the whole human race.

The pledge is often used at the beginning and end of a school term and on other special occasions.

Interpretation of the Pledge

Jamaica's national pledge is a sincere promise by the people to serve Jamaica faithfully with their skills and talents. It instils pride and commitment to service, respect for fellow citizens and promotes peace.

Protocols for use of the National Anthem

All people should stand at attention (i.e., heels together, arms straight by the sides) during the national anthem. Men should remove their hats.

On the arrival of the Governor General or Prime Minister, the first verse of the anthem must be played or sung as prescribed.

The national anthem may be sung or played at public gatherings.

The national anthem should form part of the ceremony of raising and lowering the flag at the beginning and the end of term in schools, at Independence celebrations or other special occasions. [1]

4. The National Symbols

The national symbols of Jamaica are:

A. The ackee

The ackee plant was brought to Jamaica from West Africa in 1778. It is called other names in different Caribbean countries but is eaten mainly in Jamaica.

B. The Doctor Bird

The 'Doctor Bird' or Swallowtail Humming Bird lives only in Jamaica and is one of the most outstanding of the 320 species of humming birds.

C. Lignum Vitae

The Lignum Vitae is indigenous to Jamaica and is said to have been found here by Christopher Columbus. It is said to have medicinal properties and is called the 'Wood of Life'. The wood is used for propeller shaft **bearings** in ships.

[1] **Source:** *https://jis.gov.jm/information/anthem-pledge/*

D. The Blue Mahoe

The Blue Mahoe is indigenous to Jamaica. It is used to repopulate forests affected by deforestation. The lumber from the Blue Mahoe is used for making cabinets or furniture. It is valued as a primary timber for making money.

Protocols for the National Symbols

- Treat the national symbols with respect.
- Depict them respectfully in artistic creations and commercial activities.

Acknowledging and Maintaining Pride in our National Symbols

Jamaicans have a responsibility to acknowledge and maintain pride in the national symbols. We respect these symbols as important reminders of the nation's goals, values and history. It is very important to obey the rules and guidelines for using and representing the symbols.[2]

[2] **Sources:**
https://nlj.gov.jm/jamaican-national-symbols/
https://jis.gov.jm/information/symbols/

TARGET

Addressing Individuals in Public Office

Position	Titles
Governor General	His or Her Excellency The Most Honourable
Prime Minister	If the Prime Minister is a member of the Privy Council: Right Honourable If awarded Order of the Nation: Most Honourable
Minister of Government	The Honourable
Senator	Senator Senators who are members of the Order of Jamaica or ministers of

Position	Titles
	government are addressed as: Senator The Honourable
Mayor	His or Her Worship The Mayor
Councillor	Councillor

The Governor General

The Governor General of Jamaica is the King of England's representative. Therefore, he is the head of state. The Governor General takes priority, except when His Majesty the King or a member of the royal family is present in Jamaica. The governor and his wife should be referred to as 'Your Excellency' (singular) or 'Your Excellencies' (plural), when spoken to directly. In their absence, they are referred to as 'His Excellency' or 'Her Excellency', individually or 'Their Excellencies' collectively.

Source: *https://jis.gov.jm/governor-general-delivers-throne-speech-for-new-fiscal-year/*

Figure 7: *Governor General, His Excellency the Most Hon. Sir Patrick Allen, inspects the Guard of Honour mounted by members of the Jamaica Defence Force, at the 2020/21 ceremonial opening of Parliament, (February 11).*

Protocols for the Arrival of the Governor General

- All attendees take their seats before the Governor General arrives.
- Play the national anthem as soon as the Governor General stands in his assigned place.
- Everyone attending the function stands when the Governor General enters the room and remains standing for the anthem.

The Prime Minister

Prime Ministers who are members of the Privy Council are called 'the Right Honourable (first and last name), PC, MP'. However, if the Order of the Nation has been bestowed on the Prime Minister, he/she is referred to as 'the Most Honourable'. In addition, if he/she becomes a member of the British Privy Council, the letters PC must be placed after his/her name. Therefore, 'Most Honourable' replaces 'Right Honourable' as in this example: The Most Honourable Andrew Holness, ON, PC, QC.

Figure 8: *Prime Minister, the Most Hon. Andrew Holness (fourth left); and his wife, Member of Parliament for St. Andrew East Rural, the Most Hon. Juliet Holness (fifth left), lead government ministers into Gordon House at the ceremonial opening of parliament.*

Photo by *Yhomo Hutchinson.*

Protocols for the Arrival of the Prime Minister

At formal official functions, where the Governor General is not present, the national anthem should be played to announce the arrival of the Prime Minister. Everyone present should stand at attention.

Imagine you were alive in the 1960s. Do you believe gaining independence was an exciting time for Jamaica? Part of being independent was creating our artefacts. What did our ancestors do then? Did you say create symbols and emblems that tell others who we are? What about pledges that will guide the people living in Jamaica? Can you see our ancestors asking themselves, 'How will other countries and people globally recognise Jamaica? How do we create symbols that are unique to us?'
Would you have liked to help decide what would be a national dish, bird and flower?
The journey continues. Mek wi apply wi knowledge. Let's go. Let's prep for life and PEP!

TERM 2, UNIT 1

The Physical Environment and Its Impact on Human Activities

Key Concepts to Know and Apply

- mountain
- mountain range
- hill
- valley
- plateau
- landform
- plains
- forest reserve
- summit
- slope

You will find the definitions in the glossary.

Thematic Map of the Major Mountains in Jamaica

A thematic map shows the pattern of a particular topic or theme in a geographical region, such as agricultural or mining areas, or **forest reserves**. Physical maps show the height of the land. The map below shows some of Jamaica's major **mountains**.

Source: *https://www.freeworldmaps.net/centralamerica/jamaica/jamaica-map-physical.jpg*

Figure 9*: Physical map of Jamaica.*

Criteria for Ranking Mountains and Mountain Ranges

Mountains are ranked based on their height. Two notable highlands in Jamaica are the Blue Mountains and the Cockpit Country. The Blue **Mountain Range** is located in eastern Jamaica and spans across St Andrew, St Thomas and Portland. The Blue Mountains are protected forests and boast the highest point on the island. The John Crow Mountains are to be found just east of the Blue Mountains.

The Cockpit Country spans much of southern Trelawny and parts of St James. This highland area comprises many steep-sided, rounded **hills** with depressions among them. This area is protected, so mining or large-scale commercial agriculture is not permitted.

The table below shows some of the highest points in Jamaica.

Highland area/peak	Parish	Elevation (m)
Blue Mountain Peak	St. Thomas	2256
Sir John's Peak	Portland	1930
Catherine's Peak	St Andrew	1541
Huntley	Manchester	955
Rose Hill	Manchester	845

TARGET

How Mountains Affect Weather and Climate

Highlands, such as hills and mountains, impact the weather and climate. It is usually cooler in the mountains than on **plains**. Kingston is on the Liguanea Plains, while much of Manchester sits on a vast limestone **plateau**. The average temperature in Manchester is cooler than the average temperature in Kingston.

Mountains also influence the rainfall patterns in an area. The Blue Mountain Range significantly affects the rainfall in Portland and Kingston. Trade winds approaching the Blue Mountains from the north coast in Portland are forced to rise, with the mountains acting as a barrier. When air rises, it cools and condenses to form clouds. As a result, Portland receives the highest annual rainfall on the island, as it is on the **Windward** side of the Blue Mountains.

After releasing much of its moisture in Portland, dry air mass descends towards Kingston. This results in significantly less rainfall for the parish, which is on the **Leeward** side of the Blue Mountains.

Economic Activities in Mountain Environments

Mountains facilitate numerous economic activities. Coffee is grown in the Blue Mountains and other highlands and it is mainly grown for export. The Cockpit Country region of Trelawny, however, is known for largescale yam cultivation.

Eco-tourism is expanding with attractions such as Mystic Mountain, hiking, camping and bird watching. The activities and attractions are centred around the landscape and flora (plant life) that the mountains provide.

Importance of Mountain/Hll Environments

The following diagram highlights the importance of mountains.

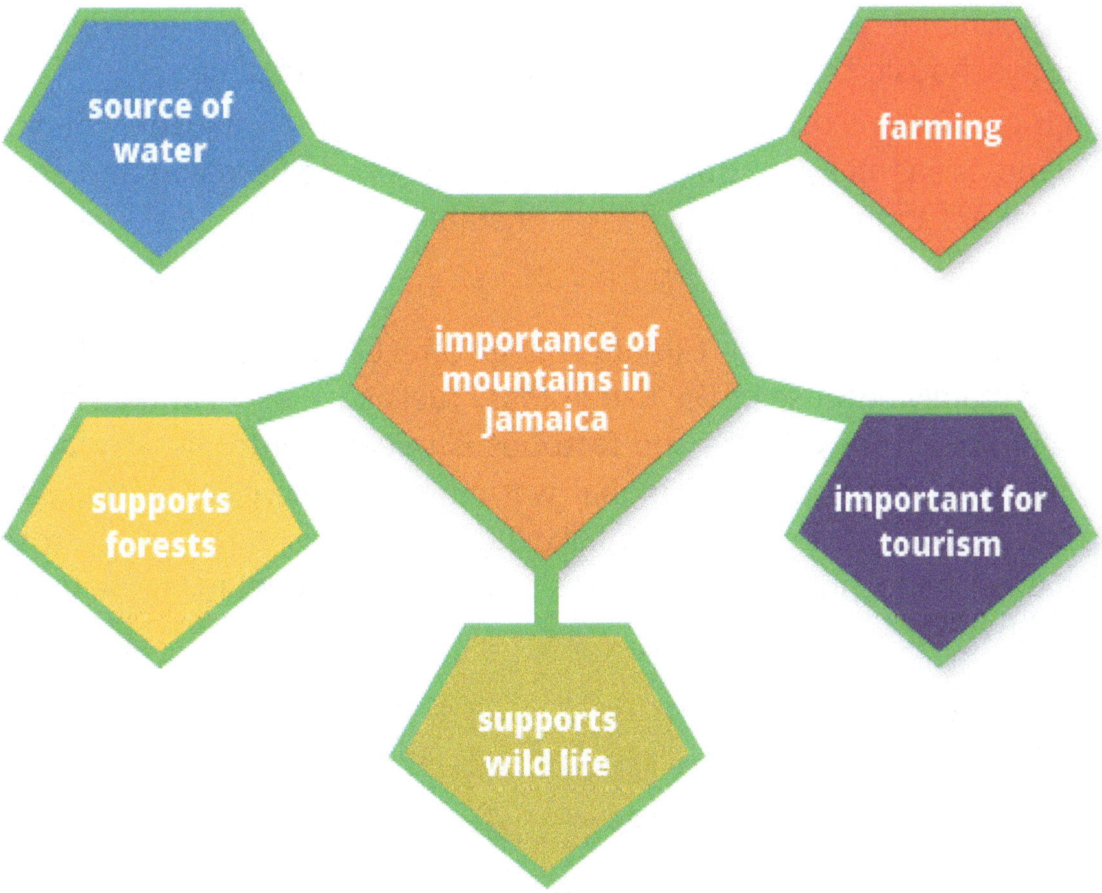

The Effects of Human Activities on Mountains

Human activities threaten and even degrade the highland areas. The following table highlights how some human activities can negatively impact highland areas.

Human Activity	Negative Impact
Agriculture: The land on **slopes** is sometimes cleared to make way for crops such as coffee or bananas.	Agricultural activities can lead to soil erosion, pollution and siltation in streams or rivers that originate from mountains.
Construction: Trees are cleared and mountains are cut to make highways, housing or settlements.	Removing trees and cutting into mountain slopes can lead to slope instability or landslides.
Logging: Trees that grow in highland areas are cut down for charcoal or furniture-making.	Logging can cause habitat destruction, soil erosion and slope instability.

Human activities resulting in climate change affect mountains. Our climate is changing mainly because of the use of fossil fuels such as oil and gas. Excess carbon dioxide from vehicles and factories is released into the atmosphere, trapping heat, which leads to global warming. The increase in the Earth's temperature leads to increased rainfall in highland areas, causing soil erosion. On the other hand, prolonged droughts can affect the vegetation and the flow of rivers in mountains.

Best Practices for Human Activities in Mountain/Hill Environments

Highland areas are an important natural resource for the country. It is therefore important to preserve them. Trees are an important feature in the ecosystem of highlands. Trees provide homes and food for animals and they

also hold the soil together. Replanting trees will reduce soil erosion and increase slope stability.
Classifying some highland areas as forest reserves can decrease destructive human activities such as mining, logging and commercial agriculture.

Public education programmes can help to inform citizens of the importance of mountains in our daily lives. When citizens are aware, they will be more likely to engage in activities that will protect and preserve highland areas.

Major Mountain Ranges of the World

The Blue Mountain Peak is the tallest mountain in Jamaica, but when compared to others around the world, it is very low.

The following table shows the tallest mountains and mountain ranges on each continent.

Continent	Country	Main Mountain Range	Highest Peak	Height in Matres
Asia	Nepal/China border	Himalayas	Mt. Everest	8,848
South America	Argentina	Andes	Aconcagua	6,962
North America	USA (Alaska)	Rocky Mountains	Mt. Mckinley/ Denali	6,190
Africa	Tanzania	East African Highlands	Mt. Kilimanjaro	5,895
Antarctica		Sentinel Range	Vinson Massif	4,892
Europe	France/Italy	Alps	Mont Blanc	4,809
Australia	Australia	Great Dividing Range	Mt. Kosciuszko	2,228

World continents and mountains

Figure 10: *Continents and major mountains of the world: The Seven Summits.*

Caring for the Environment

Now that you know many things about protecting mountains, you can take steps to become champions for the environment. You are not too young to tell others about the importance of mountains to the environment. Here are some activities that you can do to ensure that highland areas and the environment are preserved.

- Create posters to share with schoolmates and the community about the importance of mountains.
- Organise, with the help of your teachers, tree planting projects in highland regions where trees have been cut down.
- Form an environmental protection club in your school to get other students involved in fighting for the environment.
- Request a session at the next school's PTA meeting to make a presentation to parents and teachers about the importance of protecting highland regions.

TERM 2, UNIT 2

The Physical Environment and Its Impact on Human Activities

Focus question: 'How can we classify the landmasses and water bodies of the world?'

Key Concepts to Learn and Apply

TARGET

- continent
- island
- ocean
- sea
- lake
- river
- bay
- gulf
- peninsula
- isthmus
- archipelago

You will find the definitions in the glossary.

TARGET

Grid, Latitude, Longitude, Great Circle, Hemisphere

A grid on a map is used to determine the position of a place on Earth. It uses two coordinates, latitude and longitude. Latitude is an angular measurement north or south of the Equator. For example, 30 degrees north means a position that is 30 degrees north of the Equator. Lines connecting all the same latitude points are referred to as parallel since the lines are straight, run in the same direction and do not intersect or meet each other. Latitudes run from west to east.

The most important lines of latitude are:

- The Equator, which is 0 degrees. It divides the Earth into two equal parts called hemispheres
- Tropic of Cancer, 23.5 degrees north
- Tropic of Capricorn, 23.5 degrees south
- Arctic Circle, 66.5 degrees north
- Antarctic Circle, 66.5 degrees south
- North Pole, 90 degrees north
- South Pole, 90 degrees south.

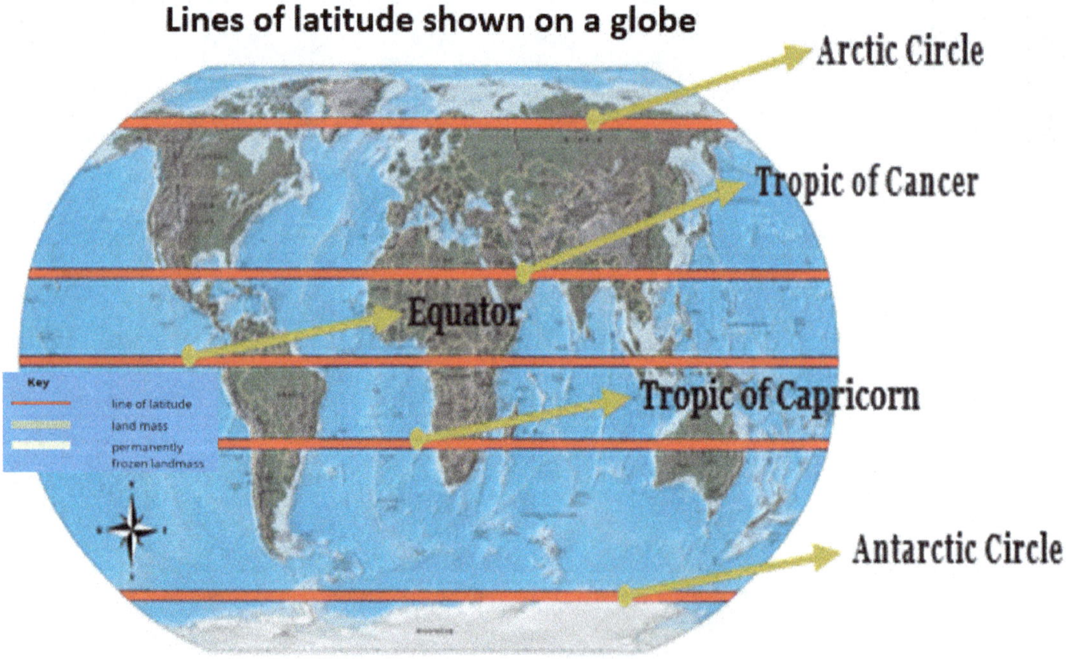

Figure 11: *Major lines of latitude.*

Longitude is the measurement east and west of the Prime Meridian and is often measured in degrees. The lines joining the same points of longitude are known as **meridians**. The **Prime Meridian** is the main line of longitude. It is also called the **Greenwich Meridian**. It measures 0°

Earth, therefore, has four hemispheres:

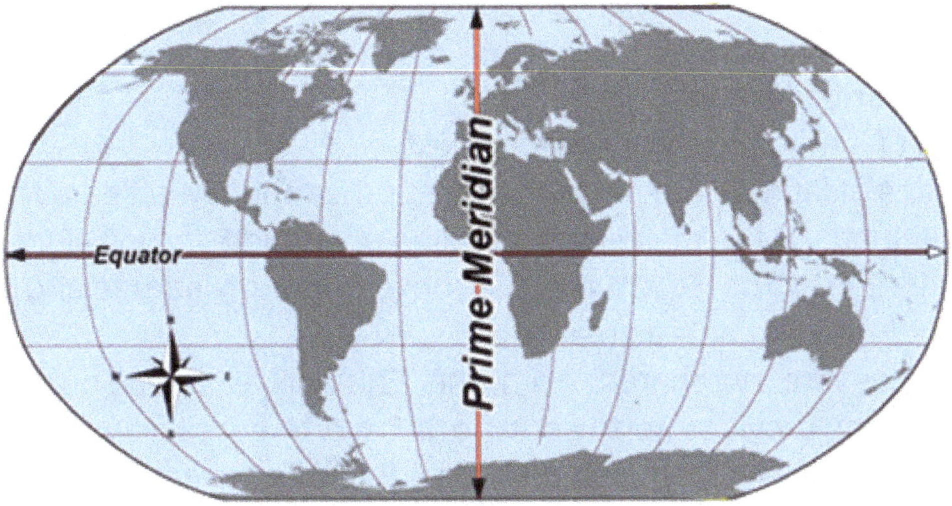

Figure 12: *World hemispheres.*

The Division of Earth by the Equator and the Prime Meridian

When you put together latitude and longitude, you get the geographic grid system that helps you to determine the exact location of a place on the Earth.

Where in the world is Jamaica? Use a map or globe to find it. Draw a circle around Jamaica.

Is it on any of the continents?

Characteristics of Latitudes and Longitudes

Lines of Latitude

a. Lines of latitude run from east to west.
b. All lines of latitude are parallel to each other and are called parallels.
c. The Equator is the most important line of latitude. It runs along the centre of the Earth. The Equator is the longest line of latitude,and is the only latitudinal line that is a great circle.
d. Latitudes are numbered from the Equator at the centre at 0^0. This numbering provides north-south location references from the Equator.
e. Among the lines of latitude, only the Equator is a great circle.

Lines of Longitude

a. Lines of Longitude run from north to south.
b. They are not parallel to each other and are called meridians.
c. They are numbered from the Prime Meridian at the centre at 0^0. This numbering provides east-west location references from the Prime Meridian.
d. All the lines of longitude are great circles.

The Great Circle

A great circle refers to the largest possible circle that can be drawn around a sphere. All meridians are circles. The Equator is the only line of latitude that is a great circle.

Great circles are important in planning routes. The shortest route between two places on the surface of a sphere is usually a great circle segment. Pilots need to identify great circles to find the shortest distance between two positions. For example, if you flew from Kingston, Jamaica, to Bangkok, Thailand, you could fly almost along the route of one of the Earth's great circles, which would be the shortest distance between those two countries.

Classification of Landmasses

Landmasses are classified as **continents** and **islands**. Continents are the largest landmasses. These continuous masses of land are bordered by bodies of salt water; continents are usually made up of various countries.

There are seven continents in the world.

1. Asia
2. Africa
3. North America
4. South America
5. Antarctica
6. Europe
7. Australia

The list you just read has been ordered from largest to smallest.

Islands are the smallest landmasses. A group or chain of islands make up an **archipelago**. Examples of islands are The Bahamas, Hawaii and Japan, which are all archipelagos.

Peninsula is a piece of land that is long and extends into a body of water. It is connected to the main land on one side , with three (3) sides surrounded by water. Examples of peninsulas are the Palisadoes Strip in Jamaica (on which the Norman Manley International Airport is located). Also the US state of Florida.

An Isthmus is a narrow piece of land that links two larger land masses usually separates two bodies of water .An example is the Isthmus of Panama which connects the continents of North and South America.

Classification of Water Bodies

The water bodies can be classified as **oceans**, **seas**, **lakes**, **rivers** and canals.

Oceans are the largest water bodies. The oceans include Atlantic, Indian, Arctic, Pacific and Southern oceans.

Seas are the second largest water bodies. Some of the major seas include the Mediterranean Sea, the Red Sea and the Caribbean Sea.

A lake is a water body that is completely surrounded by land. Two examples are Lake Superior in North America and Lake Victoria in Africa. Like lakes ,

Ponds are inland bodies of water surrounded by land. However, Ponds are usually smaller and shallower.

A river is a stream of fresh water that flows from a highland. The *source* is where a river starts, while a *mouth* is where it joins a lake, sea or an ocean. A smaller river that joins the main river is a *tributary*. Examples of rivers are the Amazon River, the Nile River and the Rio Grande.

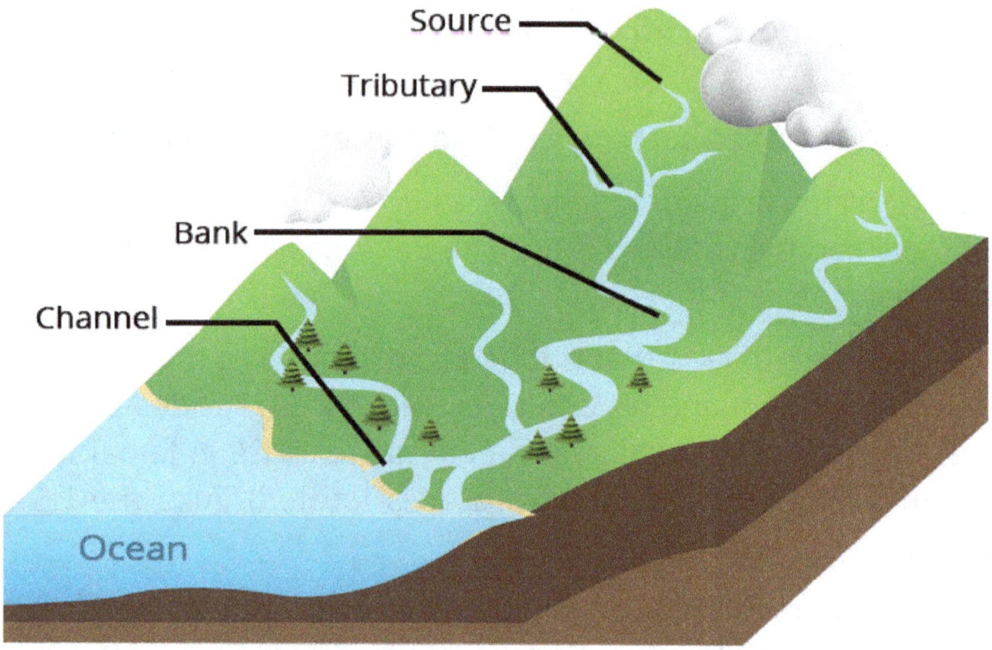

Gulfs are area of the sea or an ocean that is surrounded on three sides by land and has a narrow mouth.

Bays are areas of the sea where the land is partially surrounded by water.

Canals are usually man-made features that connect bodies of water. Examples include the Panama Canal and the Suez Canal.

TARGET

The Proportion of Landmasses to Water Bodies

Water bodies occupy almost three-quarters or 71% of the Earth's surface. The landmasses span about 29% of the Earth's surface.

The Absolute and Relative Location of Landforms and Water Bodies

We can use a relative location or an absolute location to determine a place's global address. A relative location identifies the position of a place in relation to another place. An absolute location uses the global grid system to determine a place's position using latitude and longitude coordinates.

Using Jamaica as an example, we can say its absolute location is 18 degrees north, 77 degrees west. Absolute location is always fixed. If we use Jamaica's relative location, we can say Jamaica is south of Cuba, east of Belize and north-east of Panama.

We can also use relative and absolute locations to determine the position of water bodies. For example, the Caribbean Sea's absolute location is 14.5^0 north, 74.9^0 west. It is to the north of South America, according to relative location.

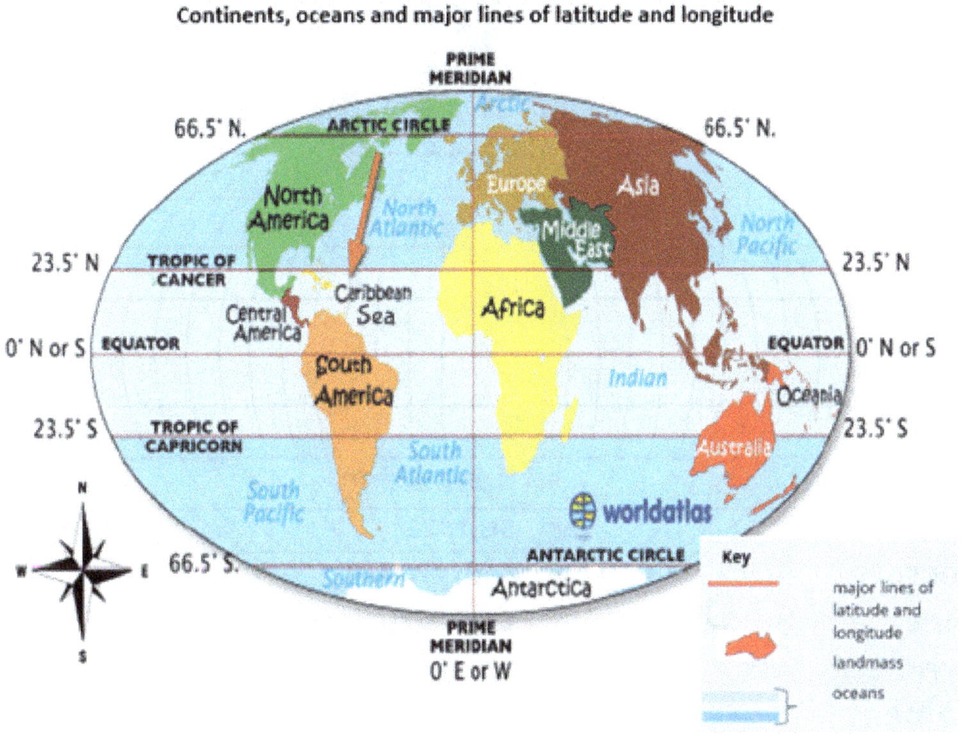

Source: *www.worldmap.com*

Figure 13: *Continent, oceans and major lines of latitude and longitude.*

TERM 2, UNIT 3

Living Together

Focus question: How are decisions made at the national level and how do these decisions affect us?

Key concepts to know and apply

- citizen
- leader
- democracy
- cabinet
- government
- parliament
- opposition
- senate
- monarch
- constitution
- vote
- constituency

You will find the definitions in the glossary.

TARGET

Requirements for Jamaican Citizenship

A **citizen** is a legal member of a country entitled to all the rights and privileges and shares the responsibilities of that country.

There are different ways to qualify for Jamaican citizenship.

1. **Birth:** Any person born in Jamaica or born outside Jamaica of Jamaican parents have an automatic right to Jamaican citizenship (Chapter two of the Jamaican constitution)
2. **Descent:** Any person born overseas of Jamaican parents can register as a citizen of Jamaica.
3. **Adoption:** Any person from another country who has been legally adopted by Jamaican parents is a citizen.
4. **Marriage:** Being married to a Jamaican citizen presents the opportunity for an adult of another country to register to become a Jamaican citizen.
5. **Naturalisation**: A citizen of another country can become a citizen of Jamaica. The person must live in Jamaica for a minimum of five years and be of good standing with respect to character and financial stability.

Source: *https://jis.gov.jm/33-foreign-nationals-granted-jamaican-citizenship.*

Figure 14: *Newly naturalised citizen received a certificate from National Security Minister, Hon. Horace Chang.*

Citizens' Rights and Responsibilities

All citizens have rights or things they are entitled to or should be allowed to access. These rights are not affected by a citizen's gender, race, religion or finances.

As a citizen of Jamaica, you have the following rights and privileges:

1. The right to life and freedom
2. The right to freedom of speech
3. The right to be treated fairly despite your colour, race or religion
4. The right to go to freedom of worship and movement
5. The rights to liberty, security and education
6. The right to **vote** in an election once you are at least 18 years old
7. The right to privacy and family life
8. The right to enjoyment of property
9. The right to protection from arbitrary arrest and detention.
10. The right to be treated humanely if you get in trouble with the law. It is your right to have a fair trial and an attorney

Source: *https://jis.gov.jm/jamaica-welcomes-32-new-citizens/*

Figure 15: *National Security Minister, Hon. Dr. Horace Chang (centre), presents Sangeeta Sharma (left) with her Registration of Jamaican Citizenship at a ceremony conferring Jamaican citizenship on 32 persons at the Police Officers' Club in Kingston. At right is Chief Executive Officer (CEO) of the Passport, Immigration and Citizenship Agency (PICA), Andrew Wynter. Photograph by Michael Sloley.*

The right of a citizen gives that citizen a set of responsibilities. As a good citizen, there are some responsibilities we have in our country:

1. To take care of public properties and participate in activities that develop their community
2. To pay our taxes
3. To serve as jurors in court
4. To report crimes we witness
5. To respect the lives and property of others
6. To be helpful and respectful to our neighbours.

This table shows some responsibilities which are linked to specific rights.

Rights	Responsibility
Liberty and security	to take care of public properties and participate in activities that develop their community
Security, liberty, education	to pay our taxes
Fair trial	to serve as jurors in court
Freedom of speech and security	to report crimes witnessed
Privacy and enjoyment of property	to respect the lives and property of others
Freedom	to respect your neighbours' rights.
Vote	To participate in the democratic process

Organisational Structure of the Jamaican Government

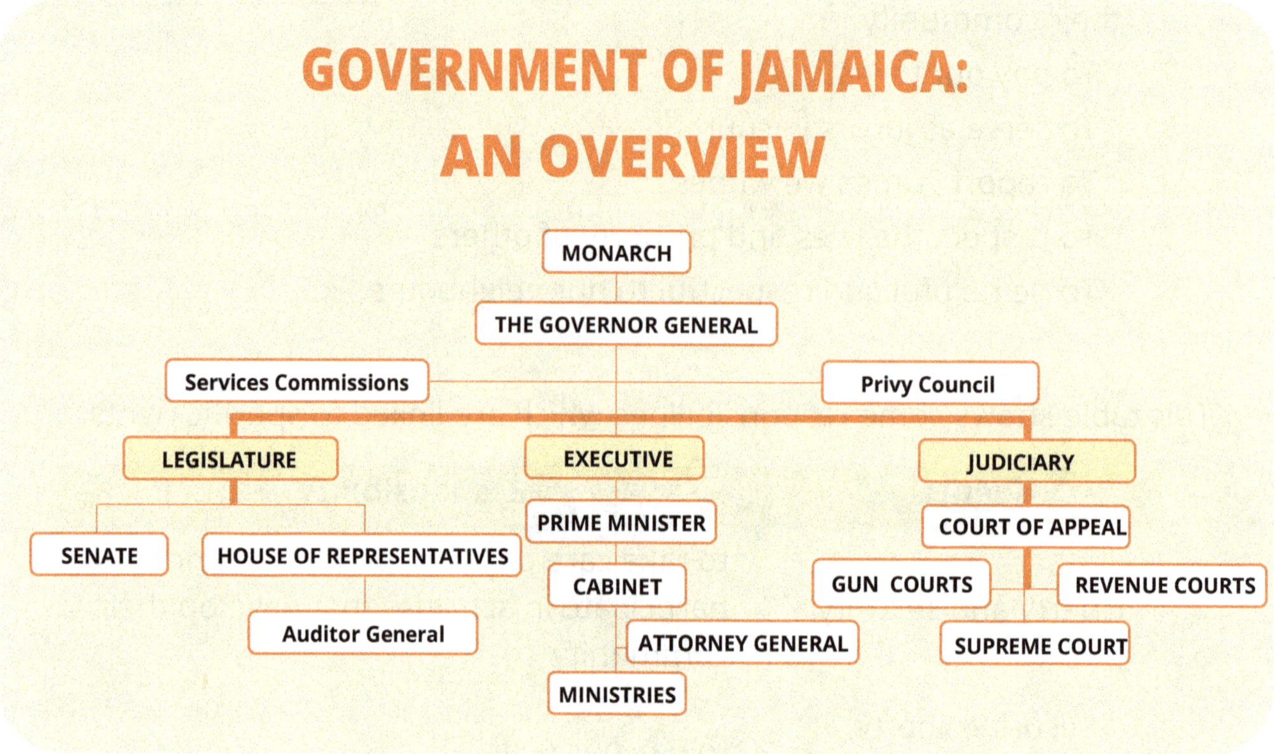

Source: *https://jis.gov.jm/features/overview-government-jamaica/*

Figure 16: *Hierarchy diagram showing the structure of the Jamaican government.*

The Jamaican **government** is headed by the **British Monarch**, King Charles III, who is represented by the Governor General. The Jamaican government is divided into three branches:

1. **The Legislature:** Known as the **parliament**, this is the law-making body of government. It is made up of two chambers or houses - the **Senate** (the upper house – 21 senators) and the **House of Representatives** (the lower house – 63 **Members of Parliament - MPs**) with the **Auditor General** reporting directly to the house of representatives.
2. **The Executive:** This branch is the policy-making body. It is made up of:
   I. The **Prime Minister**: **leader** of the country

II. The **Cabinet** – Consists of the Prime Minister and selected Ministers of government
III. The **Attorney General** – Advises the executive body on legal matters.
IV. The **Ministries** – Each Ministry is led by a Minister who has portfolio responsibilities for a specific area such as health, security, agriculture, education, industries, tourism, justice, finance, sports, foreign affairs and trade.

3. **The Judiciary:** This branch upholds the laws of the country. It includes the Court of Appeal, Supreme Court and the Gun Court. The Supreme court has authority over the following lower courts:

Night Court	Traffic Court	Drug Court	Coroners' Court
Parish Court	Family Court	Juvenile Court	Small Claims Court

Appointment of Government Officials

Source: *https://jis.gov.jm/speeches/independence-day-message-from-the-governor-general-his- excellency-the-most-hon-sir-patrick-allen-on-gcmg-cd-kst-j/*

Figure 17: *The Governor General His Excellency The Most Hon. Sir Patrick Allen ON, GCMG, CD, KSt.J.*

The Governor General is the official head of state. He represents the British **Monarch** in Jamaica and is appointed by the king (head of the monarchy in Britain) under the advice of the Prime Minister. The Governor General is the head of government, but does not belong to any one political party. His duties and responsibilities include:

1. Representing the head of state (the king) in ceremonies
2. Appointing the Prime Minister, Deputy Prime Minister and Leader of the **Opposition**
3. Appointing the members of the Cabinet, Privy Council, Services Commission, the Chief Justice and other members of the justice system on the advice of the Prime Minister
4. Granting pardon to convicts on the advice of the Privy Council
5. Giving the formal agreement to bills before they become law.
6. Deciding on recommendations for pensions of public servants and receiving credentials of newly-appointed ambassadors.

Source: *https://jis.gov.jm/speeches/new-years-message-from-the-most-hon-andrew-holness-on-mp- prime-minister.*

Figure 18: *The Most Hon. Andrew Holness, ON, PC, MP, Prime Minister of Jamaica.*

The Prime Minister is an elected representative and the leader of the governing political party. This makes him/her the leader of the country. He/She is the leader of the Cabinet and, together, they guide government policies that are used to run the country.

Other responsibilities of the Prime Minister include:

1. Advising the monarch (king) on who to appoint as the Governor General
2. Directing the Governor General on who to appoint as members of the Privy Council and the chief justice

3. Directing government activities so that they benefit the people of the country
4. Appointing ministers responsible for different ministries in the country and ensuring that they work for the good of the people
5. Selecting 13 of the 21 members of the senate.
6. Serving as the minister of defence. He is responsible for the armed forces. He also represents the country at important national meetings.

TARGET

Qualities of an Effective Leader

We need good leaders to help us make important decisions as a country. What makes a good and effective leader? Here are some of the skills and qualities of a good leader:

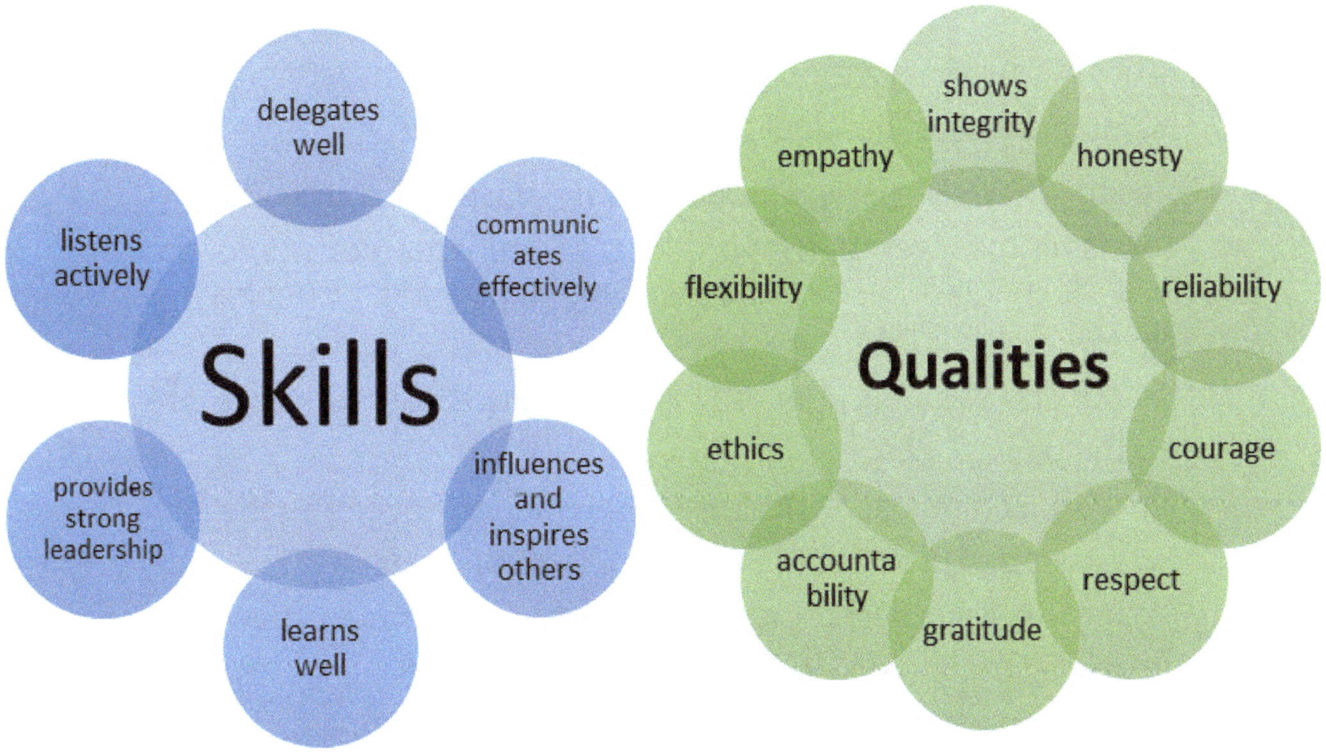

Figure 19: Qualities and skills of an effective leader

Main Goods and Services Provided by the Government

The government of the country is responsible for providing a number of goods and services for its citizens. Different ministries are responsible for ensuring that these goods and services are provided and available to all citizens. Here is a list of ministries and their responsibilities:

1. Education and Youth– public education, youth coordination and provision
2. Local Government – garbage disposal, fire protection, street lighting, water, recreational centres and national parks.
3. Culture, Gender, Entertainment and Sport – culture and creative entertainment, national celebrations (independence, heritage), sports
4. National Security and Justice – police services and national defence
5. Transport and Mining – public transportation, road safety, air and sea transport
6. Economic Growth and Job Creation – job creation, investment, promotion, building factories, developing small businesses
7. Health and Wellness – public health care, hospitals and clinics
8. Tourism – manages tourism products and services (attractions, hotels)
9. Science,Energy and Technology – energy supply and technology development
10. Labour and Social Security – safety and health at the workplace, protection for the poor
11. Industry, Commerce, Agriculture and Fisheries – oversees the agricultural industries and provides avenues for export of products
12. Finance– collects government revenue (taxes) and plans how to spend it, creates a strong economy for the country.
13. Justice – ensures the rule of law is upheld, protects human and animal rights and keeps the country safe.
14. Constitutional Affairs – reforms laws and reviews the **constitution**.

Government earns revenue or money to provide goods and services for the citizens through taxes. These include:

i. **Property taxes:** taxes paid by landowners
ii. **Income tax:** taxes paid by employees as a percentage of their wages
iii. **Sales taxes:** taxes paid on the purchase of some goods and services.

How Decisions Are Made in the Government, School and Home

Government

A good leader must make decisions for the good of all the people. However, all decisions have intended (planned) and unintended (unplanned) impacts on the citizens. The Cabinet, with the Prime Minister as head, is the policy-making body of the country.

Here are five steps taken by the government in making decisions or creating policies:

1. Develop a concept: this gives information on the problem and how it will be solved.
2. Prepare and discuss the policy/plan and identify persons who are responsible for the plan.
3. Have public consultation: the citizens must be aware of the plans, help to formulate them and be conscious of the effects these plans can have on them.
4. Amend (revise or change) policies/plans to address the concerns of the public.Have the approval of the Cabinet for the final plan.

Schools

The Ministry of Education and Youth is the head of public schools in Jamaica. Next in authority is the chairman of the board, followed by the principal. Below the principal are vice-principals, then the senior teachers and then the clerical administrators. Below this level are the other teachers, followed by the students.

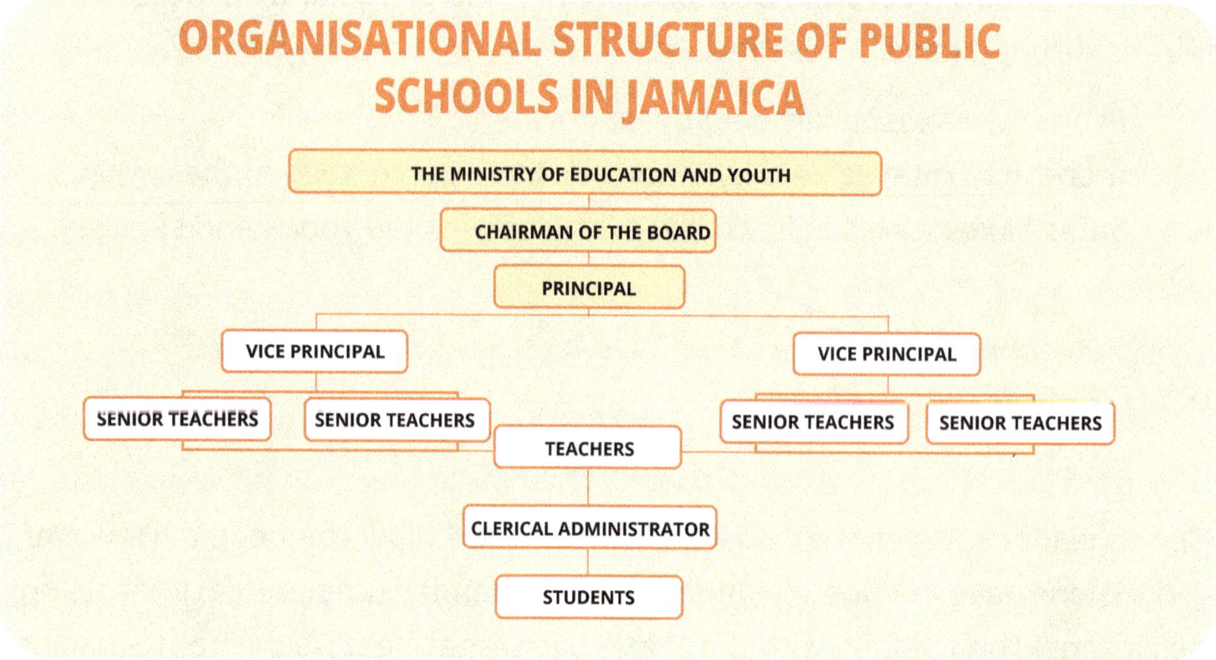

Figure 19: *Organisational structure of some positions for Jamaica's public schools. This structure is apart of how decisions are made at schools as oppose to in the government.*

Home

in a home, there is also a structure of how decisions are made. Some households are led by a parent, guardian or an elder in the household. When a problem or objective is identified, the head(s) of the home would discuss the problem and come to a decision.

The children and other dependent members of the home may then be informed of the problem and the decision and their role in the solution process.' In many homes there is a growing policy of consultation with the children for the setting of rules for chores, and discipline, especially in the prepuberty and adolescent stages.

Justice and Fairness for Citizens

The Jamaican justice system is patterned after the British system of law. This system includes the petty sessions that deal with less offensive matters. Each parish has a Parish Court with power to hear civil and criminal matters. The Parish Court has a number of divisions: Family, Coroners, Tax, Juvenile, Traffic. Small Claims, Night, Drug, and Gun Courts. Severe crimes such as rape, treason, and murder are not tried by the Parish Courts, but are referred to the Supreme Court after a preliminary hearing.

Each parish also has Petty Session Courts to hear minor criminal matters, such as resisting arrest. Justices of the Peace serve as judges in the Petty Sessions. Petty Session Courts require a minimum of two Justices of the Peace, to be properly constituted.

https://en.m.wikipedia.org/wiki/Judiciary_of_Jamaica#:~:text=The%20Petty%20Sessions%20hear%20minor,Peace%2C%20to%20be%20properly%20constituted

The Supreme Court is responsible for dealing with very serious criminal, civil and constitutional matters in the country. If a citizen feels that his matter was not treated fairly in the local courts, he can appeal to the Caribbean Court of Justice (CCJ) and, finally, to the Privy Council in England.

The duties and responsibilities of the Jamaica Constabulary Force (the police) are to serve, protect and reassure all people by being fair and not discriminating when enforcing the law. They should protect the life and property of all citizens and prevent crime, while keeping the peace in the community. All citizens have the right to a fair trial if they are suspected of breaking the law.

A Justice of the Peace (JP) works within individual communities to promote and protect the rights of citizens. They are usually selected on the grounds that they are persons living in their communities with sound integrity. The JP works through the Public Law Restorative and Preventative Justice Unit of the Ministry of Justice (MOJ).

That was a lot of information. Was most of it familiar to you? Are you ready to apply your knowledge? Come journey wid mi.
Let's prep for PEP and life.
SCHEDULE ROOSTER CROWING

Unit Assessment Based on the NSC

Term 1, Unit 1

Focus question: How can we promote and preserve our Caribbean culture?

'Preservation of one's own culture does not require contempt or disrespect for other cultures.'

– *Cesar Chavez*

TERM 1, UNIT 1

Promoting and Preserving our Caribbean Culture

Now that you have read and understood the key concepts, we will now need to assess how much you have learned. Read and complete the columns below by placing a tick (✓) in the column that best describes your knowledge of the concepts. You may choose:

- ✓ *I understand*
- ✓ *I somewhat understand*
- ✓ *I don't understand.*

When you are finished, ask your teacher to review with you before you prepare for the next unit. Good luck!

I need to be able to:	Column A I completed the following:	Column B How did I do?		
		1. I understand.	2. I somewhat understand.	3. I don't understand.
Develop working definitions for and use the following: ▪ indentured servant ▪ indentureship ▪ contract ▪ festival ▪ carnival ▪ immigrant ▪ migration ▪ push and pull factors.				
Recall the meaning of culture, heritage, ethnic group.				
Gather information and use mathematical skills to construct timeline showing the arrival of the various ethnic groups to the Caribbean.				

I need to be able to:	Column A I completed the following:	Column B How did I do?		
		1. I understand.	2. I somewhat understand.	3. I don't understand.
Create a thematic map of the world showing the places of origin of the ethnic groups that came to the Caribbean.				
Distinguish between the pull and push factors that led to migration of the East Indians and the Chinese to the Caribbean.				
Describe the life of Chinese and East Indian immigrants on the plantation from the 19th to the 20th century.				
Describe the relationship between East Indians, Chinese, Europeans and Africans in the post-emancipation period.				
Discuss the contribution of the East Indians and Chinese to the Jamaican economy.				
Categorise aspects of culture as goods and services.				
Describe and compare the traditions and celebrations of different ethnic groups by examining the following: Diwali (Divali)/ Hosay, Easter/Christmas, Crop Over, Chinese New Year.				
Identify various ways of preserving Caribbean culture.				
Describe various strategies that are used to promote Caribbean culture.				
Create goods (products) and services (strategies) to promote Caribbean culture.				
Value the contribution of the East Indians and Chinese to Caribbean culture.				
Be aware of the economic value of culture and creative industries.				

Objective: Develop working definitions for and use the following: indentured servant, indentureship, contract, festival, carnival, immigrant, migration, push and pull factors.

Item Types: Order match and single selected responses

1. The table below contains a list of concepts and matching definitions. In the concept column, write the letter for the corresponding definitions on the lines given. Each word should be used only once.

Concepts	Definition
___ indentured servant	**A.** an event which includes entertainment, celebrations and processions
___ indentureship	**B.** a legally binding agreement between parties
___ contract	**C.** a special day or period of celebration, usually in memory of a religious event, with its own social activities, food or ceremonies
___ festival	**D.** a person who leaves his/her country and lives in a foreign country
___ carnival	**E.** a condition or situation that encourages individuals or groups to migrate
___ immigrant	**F.** movement from one location to another area to settle
___ migration	**G.** a benefit that attracts a person or group to a place
___ push factor	**H.** a paid immigrant worker who is bound by a contract for a number of years
___ pull factor	**I.** an economic system in which workers from other countries are contracted to work for a period

Read the paragraph carefully. Choose the option with the set of responses which accurately completes the paragraph. Each word can only be used once.

indentured servant	festival	migration
indentureship	carnival	push factor
contract	immigrant	pull factor

Many groups of immigrants came to Jamaica once slavery ended. Some of the conditions, like drought and lack of freedom, were_______________(s) that caused people to leave their homes for what they believed was a better place. One benefit that drew many immigrants to Jamaica was the availability of jobs. Some groups, like the Chinese and East Indians, signed a____________ and worked as part of the________________ programme, an economic system that allowed workers to become___________________(s). They could work for a number of years and then earn their freedom. Immigrant groups from various cultures remain a permanent part of Jamaica's diverse population. There are a number of ______ and ________ events that help to celebrate the different groups' contributions to Jamaica's history.

2. Which of the following is the correct choice for the passage above?

A. migration, indentured servant, immigrant, pull factor, push factor, indentureship, contract, secular, religious
B. push factor, contract, indentureship, indentured servant, secular, religious
C. indentured servant, immigrant, migration, pull factor, push factor, contract, indentureship, secular, religious
D. migration, immigrant, pull factor, push factor, indentureship, indentured servant, contract, secular, religious.

Objective: Recall the meaning of culture, heritage, ethnic group.

Item Type: Single selected response

The definitions in the source were taken from a glossary. Use them to answer items 3 and 4.

Definition 1	Definition 2	Definition 3
Heritage is often used to describe things of value, traditions, or artefacts, passed down through generations.	Culture involves the customs, ideas and behaviour of a particular group of people.	An ethnic group shares similar history, characteristics such as skin colour, language, religion, traditions and customs.

3. Which statement BEST describes the relationship between heritage, culture and ethnic group?
   A. An ethnic group consists of people who share the same heritage and culture.
   B. A person's heritage and culture are not related to the ethnic group they came from.
   C. Culture is not important in describing a person's heritage or ethnic group.
   D. An ethnic group is not related to a person's culture or heritage.

4. Which term refers to Jamaica's collection of pottery, wooden relics and people's names?
   A. Ethnic group
   B. Culture
   C. Heritage
   D. None of the above

Objective: Gather information and use mathematical skills to construct timeline showing the arrival of the various ethnic groups to the Caribbean.

Item Types: Single selected, order match and short constructed responses

5A. Match the group of immigrants to their year of arrival in Jamaica.

Immigrants	Year of Arrival
Africans	1494
Jews	1513
Chinese	1854
Indians	1834
Germans	1845

5B. Use your answer from 5A to create a timeline for the arrival of the immigrants mentioned.

6. From the timeline above, how many years passed between the arrival of the first and last groups of immigrants to Jamaica?

A. 1494 years
B. 336 years
C. 360 years
D. 1849 years

Objective: Create a thematic map of the world showing the places of origin of the ethnic groups that came to the Caribbean.

Item Type: Single selected response

Examine the map below and use it to answer items 7 and 8.

7. Which ethnic groups originated from the locations indicated by X and Y above?

   A. Africans and East Indians
   B. Chinese and Germans
   C. East Indians and Chinese
   D. Africans and Germans

8. Which ethnic group come from the location on the map represented by the letter Z?

   A. African
   B. Chinese
   C. Spanish
   D. German

Objective: Distinguish between the pull and push factors that led to migration of the East Indians and the Chinese to the Caribbean.

Item Type: Table grid response

9. Categorise the factors below as having a push or pull effect on migration to Jamaica. Use a tick (✓) to indicate which type of factor each condition or event falls under.

Condition or Event	Push Factor	Pull factor
A. Restricted freedom		
B. Promise of jobs		
C. Free travel and housing		
D. Famine		

Push factors are reasons that caused the emigrants to leave their country.

Helpful tip: Think of 'push' as being away from you: an outward movement similar to when you push something out of the way and 'pull' as drawing an object closer to you.

Objective: Describe the life of Chinese and East Indian immigrants on the plantations from the 19th to the 20th century.

Item Types: Single selected and short constructed responses

Read the following passage and use it to respond to items 10 to 12.

The Chinese and East Indians started to migrate to Jamaica in the second half of the 1800s. Both groups came as indentured servants to work on the West Indian plantations after slavery ended. They came here for new opportunities to make a better life for themselves and their families.

The conditions they experienced on the plantations were not what they expected prior to arriving. They were provided with housing, food, clothing and tools for farming. However, the Chinese were very unhappy because they did not get the accommodations they were promised. Their rooms had no windows and the floors were just dirt. They had to work from sunrise to sunset for seven days with very low wages.

The East Indians also received little pay for working long hours for five to six days per week. They lived in barracks that consisted of a few small rooms. Each room housed at least one family, which made the East Indians very disappointed.

These immigrants did not have much freedom and could not leave the plantation without a permit. The regular medical check-ups that were promised rarely happened and as such, many suffered from various illnesses.

Despite the many disappointments, the immigrants had hope because the contract to which they agreed gave them some incentives which included a free passage back to their homeland.

10. Which statement BEST describes the factors that drew the East Indians and Chinese to Jamaica?
    A. Both the Chinese and East Indians came to the Caribbean hoping to find work in grocery stores.
    B. The East Indians came as indentured servants, while the Chinese were used as slaves on the sugar plantations.
    C. The Chinese and East Indians came to the Caribbean for a better life. However, both ethnic groups soon left the island permanently.
    D. Both the Chinese and East Indians immigrated to the Caribbean as contract workers, for better job opportunities and a better life, remaining even after the indentureship had ended.

11. In your own words, describe the life of the Chinese and East Indian immigrants.

____

____

____

____

____

____

12. Which statement is NOT true about the living conditions of the East Indian and Chinese immigrants?
    A. The Chinese and East Indians were given housing and clothing on their arrival.
    B. The indentured servants were very healthy and provided with good and regular health care.
    C. The Chinese and East Indians worked long hours for little pay.
    D. The indentured servants could not move freely and required permission from the authorities to leave the plantations.

Objective: Describe the relationship among East Indians, Chinese, Europeans and Africans in the post-emancipation period.

Item Types: Single selected, table grid and short constructed responses

Use the following passage to answer items 13 and 14.

Initially, the African freed persons, Indians and Chinese had strained relationships on the sugar plantations. There were conflicts among them that often developed into fights. The formerly enslaved Africans saw the Chinese as a threat, and they treated the East Indians as inferior because they received less pay than they – the freed Africans. The East Indians, however, saw themselves as being superior to the Africans since their complexion was lighter. This idea came from the Indian Caste System that valued lighter skin tones.

Several immigrant groups settled in Jamaica after slavery was abolished in 1838. Most of the immigrants came to the island as indentured servants, working on sugar plantations for a set amount of time.

The relationships between the Europeans and the Africans and their descendants did not improve after slavery. The white population considered themselves superior to the blacks, even if the blacks were wealthy and educated. White employers continued to treat the former;y enslaved African brutally, giving them plenty of hard work and using harsh and abusive speech to them.

The relationship among them improved as the years passed. The descendants of these groups sought to mend the broken relationship, which is evident in cultural celebrations and interracial cohabitation and marriages.

13A. After reading the passage, why do you think formerly enslaved saw the Chinese workers as threats?

____

____

____

____

____

13B. The author mentioned the following: 'The relationship among them improved as the years passed.' What do you think led to this change?

____

____

____

____

____

____

13C. Read each statement in the table below. Use a tick (✓) to show whether or not the passage you read supports it.

Statement	Supports	Does not support
A. Immigrants were always the best paid workers.		
B. The relationship among the ethnic groups was poor and resulted in many disagreements.		
C. Conflicts that developed among the immigrants were as a result of their varying customs and beliefs.		
D. Immigrants protested and fought for better work conditions.		

14. Based on the passage, what can you conclude about immigrant relationships?

   A. Immigrants treated members of other ethnic groups better than they treated their own.
   B. Some immigrants were not afraid to show their discontent.
   C. All immigrants were treated better than the formerly enslaved.
   D. Immigrants likely supported and stayed within their own ethnic groups.

Source: *SMU Central University Libraries - https://commons.wikimedia.org/w/index.php?curid=53440300*
Figure 20: *Indentured labourers on a plantation.*

Objective: Discuss the contribution of the East Indians and Chinese to the Jamaican economy.

Item Type: Single selected response

Read the information from the sources and use them to answer items 15 and 16.

nnnnnnnnn

Chinese

- Brought many food products such as sauces, rice dishes and sweet and sour meats to the Caribbean. Chinese restaurants are still popular today.
- There was a huge growth in Chinese-owned food establishments such as supermarkets, shops, businesses and wholesales.
- The Chinese have consistently grown their businesses in the manufacturing, retail and wholesale industries.

East Indians

- Brought many food items to Jamaica which were cultivated, sold and used in restaurants.
- They used their lands to do peasant farming.
- Most Indians worked in the field, which created opportunities for the freed Africans to work in factories as craftsmen and as police officers.
- The East Indians are successful business owners and professionals, such as doctors, lawyers and teachers.
- They are involved in manufacturing and selling wholesale and retail, often employing many workers.

15. Which of these contributions BEST impacts Jamaica's economy?

A. Indian music and dance traditions help Jamaican dance today.
B. The Chinese festivals are celebrated today in Jamaica.
C. Arranged marriages help promote marriages that last longer.
D. Chinese grocers impact the expanding food and beverage industry on the island.

16. Based on the sources on the previous page, the Chinese and East Indians were involved in various economic activities. Which of the statements below adequately describes the similarity in their contribution to the Jamaican economy?

A. The East Indians and Chinese owned private family businesses that made them rich.

B. The food items sold in their restaurants earned a lot of money.

C. The Chinese and East Indians hired many individuals to manufacture and sell goods that benefited the country's economy.

D. The East Indians and Chinese have made major contributions to Jamaica's culture, allowing for money to be made at festivals and other celebrations.

'Education is the passport to the future, for tomorrow belongs to those who prepare for it today.'
Malcolm X
One last set of questions for this unit. Come wid mi. Let's prep for PEP and life.

Objective: Categorise aspects of culture as goods and services.

Item Types: Table grid and single selected responses

17A. Examine the pictures and use a tick (✓) to indicate whether they are examples of goods or services.

Picture	Goods	Service
rum 		
bauxite 		
music 		
sugar 		

17B. What does Jamaica's economy primarily depend on?

A. music
B. services
C. goods and services
D. goods

18. Which of the goods and services shown (rum, reggae music, sugar and bauxite) had the greatest impact on Jamaica's economy after the abolition of slavery?

A. rum
B. reggae music
C. sugar
D. bauxite

If commercial crops, such as sugar, coffee and tobacco planted by the enslaved were so profitable, then why aren't these crops as profitable for freed men? Now that we have more freedom to plant crops and no one is beating us and forcing us to do so, why don't we have more farmers? What do you think?
SCHEDULE
ROOSTER
CROWING

Objective: Describe and compare the traditions and celebrations of different ethnic groups by examining the following: Diwali (Divali)/Hosay, Easter/Christmas, Crop Over, Chinese New Year.

Item Types: Single selected and short constructed responses

The table below shows some celebrations and traditions of the ethnic groups who came to the Caribbean. Use this table to respond to questions 19 to 21.

Country	Event	Description
India	Diwali (Divali)	This festival of lights is a nine-day celebration paying tribute to the goddess of prosperity.
India	Hosay	This three-day Muslim festival involves dancing in the streets to the sound of drums.
Jamaica	Easter/Christmas	These are Christian festivals celebrating the birth, death and resurrection of Jesus.
Barbados	Crop Over	This marks the harvesting of sugar cane. People dress up, participate in a parade and support local vendors.
China	Chinese New Year	A fifteen-day celebration that begins in early spring with the arrival of the new moon. Families get together and eat, dance and have fireworks.

19. Compare the countries' celebrations. Which of the following statements BEST outlines the importance of these traditions?

    A. Each country has its own unique celebrations that are not celebrated anywhere else.
    B. Ethnic groups share celebrations as people migrate because it is important to keep traditions alive.
    C. Each country only celebrates one tradition a year because it is the most important holiday.
    D. Ethnic groups do not allow cultures to fuse their celebrations and traditions in order to keep the indigenous traditions.

20. Which statement below does NOT describe the events that take place at Christmas for the Christians?

    A. Family members coming together to enjoy meals and exchange gifts
    B. Singing carols and attending church services
    C. Decorating homes with colourful lights and ornaments
    D. Breaking bread and drinking wine, which symbolise the body and blood of Christ

21. Choose TWO of the cultural events mentioned in the table. Discuss THREE similarities that they share.

____

____

____

____

____

____

____

____

Objective: Identify various ways of preserving Caribbean culture.

Item Types: Table grid and short constructed responses

22A. For each statement below, place a tick (✓) to indicate in the box whether or not it shows a way to preserve Caribbean culture.

Statement	Preserves	Does not preserve
A. We can preserve our Caribbean culture by learning about and celebrating our food, religion, performing arts, sports and language.		
B. We can preserve our Caribbean culture by limiting our exposure to other cultures.		
C. We can protect our landmarks, heritage sites and important buildings by not destroying or defacing them.		
D. We can preserve our culture by insisting that emigrants continue to share aspects of our culture with others in their new home: our food, music, dance and language.		

22B. Choose ONE ethnic group in Jamaica. Identify ONE significant contribution that this group has made to the Jamaican culture.

____

____

22C. Suggest ONE way of preserving the culture of the ethnic group named in (b) above.

Source: *https://jis.gov.jm/photos-miss-world-2019-at-reception.*

Figure 21: *Prime Minister, the Most Hon. Andrew Holness, with Miss World, 2019, Toni-Ann Singh (third left) Miss World 1976, Cindy Breakspeare Bent (left) and Miss World 1993, Lisa Hanna Lake at a reception held at the Office of the Prime Minister on Sunday, December 22, 2019.*

Objective: Describe various strategies that are used to promote Caribbean culture.
Item Type: Multiple selected response

Read each question carefully before you select your responses.

23. Which of the following are the TWO BEST strategies for promoting Caribbean culture?

   A. Limit access to foreign influences and values.
   B. Encourage interactions between different generations to share stories and traditions.
   C. Teach students Caribbean history in school.
   D. Provide more performing spaces for cultural and creative expressions and events.

24. Which of the following are TWO reasons for promoting Caribbean culture?

   A. Promoting Caribbean culture can boost the cultural identity of a people.
   B. Promoting Caribbean culture helps prevent tourism.
   C. Promoting Caribbean culture decreases the strength and character of its culture.
   D. Promoting Caribbean culture shows how our ancestors overcame oppression and maintained their cultural identity, strength and character.

25. Which strategies would be most helpful in promoting Caribbean culture?

   A. Sell goods to tourists and increase profits.
   B. Encourage locals and visitors to enjoy authentic cultural experiences and events.
   C. Offer more fast food and beverage options for tourists to consume.
   D. Increase the number of regional activities through sports and music.

'To merely observe your culture without contributing to it seems very close to existing as a ghost.'
- *Chuck Palahniuk*
Do you agree? What can you contribute to Jamaican culture?
Come wid mi. Our journey is just starting. Let's prep for PEP and life!
SCHEDULE
ROOSTER
CROWING

Objective: Value the contribution of the East Indians and Chinese to Caribbean culture.

Item Type: Short constructed response

Use the dialogue to answer item 26.

We are all becoming like Americans. We are wearing their uniforms.
Their uniforms? What do you mean?
Everyone wears jeans and T-shirts and jackets and ties. I miss the old days of seeing Africans in dashiki, Indians in kurta-dhoti and sarees and Chinese in their hansu.
You have a point, but our national costume is a great way to promote our culture.
Seriously, you would want to now dress like Miss Lou? Good luck with that!
SCHEDULE ROOSTER CROWING

26. Based on the dialogue between the two friends, identify ONE product or service that can promote Caribbean culture. Explain your choice.

____

____

____

____

____

____

____

____

Source: *https://jis.gov.jm/features/new-parliament-building-to-reflect-out-of-many-one-people.*

Figure 22: *The winning concept design for the new Houses of Parliament building titled 'Out of Many, One People' by esteemed architect Evan Williams and his team, Design Collaborative.*

Objective: Value the contribution of the East Indians and Chinese to Caribbean culture.

Item Types: Table grid and short constructed responses

27. Read each statement in the table below and indicate with a tick (✓) if each item brought cultural value to the Caribbean.

Statement	Cultural Value
A. The East Indians brought their festivals and celebrations like Diwali and Hosay to the Caribbean.	
B. The Chinese developed their businesses until the small grocery shops grew into large enterprises. They embraced not only retailing, but also wholesaling and other types of activities.	
C. The Indians brought their religions such as Hinduism and Islam, which are still practised.	
D. The Chinese represents a very small proportion of the Jamaican population.	
E. The Chinese brought some of their food preparation techniques to the Caribbean.	

28. Explain what might have happened if the Chinese and East Indians had not migrated to Jamaica.

____

____

____

____

____

____

Objective: Gather information from multiple sources using the origin to guide selection.

Item Types: Single selected and multiple selected responses

Read the excerpts below and then use them to answer items 29 and 30.

Source 1	Source 2	Source 3
The original inhabitants of Jamaica can be traced back to 2,500 years ago, when the Tainos settled in Jamaica. These Taino Indians were made extinct by the Spanish, starting with Christopher Columbus in 1494.	The Jamaican National Heritage Trust is an organisation responsible for preserving historical sites, such as archaeological sites, cemeteries, forts and national hero sites.	In 2017, Her Excellency the Most Honourable Lady Allen said in a speech, 'There is evidence that we are losing some of our customs, traditions, national landmarks and cultural expressions. Many of us do not respect our landmarks, as we destroy and deface them.'

29. Based on the sources of information, which TWO statements are true?

A. Some of Jamaica's original sites of settlement are gone.
B. Mining companies are destroying all the ancient places in Jamaica.
C. The government is making efforts to protect historical places all around Jamaica.
D. Some Jamaicans do not care about the history of Jamaica.

30. Which of the sources support the view that Jamaicans face challenges from within their own community when it comes to respecting their culture and history?

A. Source 1
B. Source 2
C. Source 3
D. Sources 1 and 2

Source: *https://jis.gov.jm/information/jamaica-heritage-sites/st-james-heritage-sites.*

Figure 25: *The Rose Hall Greathouse is a national heritage site. It has been preserved for its cultural value.*

Objective: Be aware of the economic value of culture and creative industries.

Item Types: Single selected and short constructed responses

Read the passage below and use it to answer items 31 and 32.

Culture and creative industries like film and television, music, advertising, fashion, performing arts and animation are good for the economy. In 2013, the industries contributed revenues of US $2,250 billion dollars globally. Jobs were created for more than 29 million people worldwide.

Jamaica is no exception and benefits from creative industries. Under the Jamaica 55 Legacy Project banner, development will be done in the areas of sports infrastructure, entertainment and culture, national monuments, gender infrastructure and Jamaica 55 publications.

The legacy projects will help preserve Jamaica's culture and creative expression and showcase the island's rich cultural diversity. The legacy projects are also expected to stimulate innovation. Donors are pledging support by providing funding to creative and cultural entrepreneurs and small businesses.

Source: *http://www.jamaicaobserver.com/opinion/cultural-and-creative-industries-finally-getting- attention_106403?profile=1096.*

31. According to the passage, which statement identifies a way that cultural and creative industries provide economic value?

    A. They help preserve Jamaica's culture and creative expression.
    B. They showcase the island's rich cultural diversity.
    C. The industries contributed revenues of more than 2000 billion dollars (US) and jobs for more than 29 million people globally.
    D. They provide film, television and music entertainment.

Economic value is the monetary benefit a country can obtain from its goods and services based on what these goods and services are worth.

32. Identify and discuss TWO ways that culture and creative industries help to build the economy.

Remember to evaluate your work by completing the table at the beginning of the unit. You can always review the concepts as well as ask for help.

I understand.

I somewhat understand.

I don't understand.

Unit Assessment Based on the NSC

Term 1, Unit 2

Focus question: 'How did Jamaica achieve its independence?'

PHOTO: JIS PHOTOGRAPHER

Dancers give an exhilarating performance to a medley of dancehall music, during the 51st Independence Grand Gala at the National Stadium in Kingston, on August 6, 2013.

CHAPTER 2: TERM 1, UNIT 2

Our Common Heritage

I need to be able to:	Column A I completed the following:	Column B How did I do?		
		1. I understand.	2. I somewhat understand.	3. I don't understand.
Develop working definitions for the following: Independence, colonial rule, commonwealth, constitution, nation, monarchy, trade union, political party, self-government, universal adult suffrage, franchise, revolution.				
Use mathematical skills to construct timeline to show major developments in Jamaica's history from arrival of the Tainos to Independence.				
Examine, compare and evaluate multiple sources that outline the life and work of Marcus Garvey, Norman Manley, Alexander Bustamante.				
Apply lessons learnt from the lives of Marcus Garvey, Norman Manley and Alexander Bustamante to new situations.				
Name the major personalities involved in the independence movement in Jamaica, Cuba, Haiti.				

I need to be able to:	Column A I completed the following:	Column B How did I do?		
		1. I understand.	2. I somewhat understand.	3. I don't understand.
Compare how Independence is commemorated in Jamaica and other countries.				
Explain the significance of Independence Day.				
Formulate questions about Jamaica's decision to pursue independence and conduct research to answer these questions.				
Distinguish between dependent and independent countries in the Caribbean.				
Show appreciation for the work done by individuals in Jamaica's Independence movement.				
Weigh the arguments for and against being an independent nation and draw conclusions about Jamaica's decision to pursue independence.				
Resolve conflicts amicably while completing tasks in collaborative groups.				
Compile and arrange alphabetically a list of sources including author, title, publisher and date of publication.				

Objective: Develop working definitions for the following: independence, colonial rule, commonwealth, constitution, nation, monarchy, trade union, political party, self-government, universal adult suffrage, franchise, revolution.

Item Types: Short constructed and single selected responses

1A. Use the following words or phrase to write a paragraph.

constitution   self-government   monarchy
colonial rule   nation

____

____

____

____

____

____

____

1B. You have been asked to research and report on a Caribbean neighbour's journey to Independence. Your report must include a glossary, which is an alphabetised list of keywords or phrases and their meanings. Use these words or phrases to create the glossary.

commonwealth   universal adult suffrage   trade union
franchise   political party   revolution

____

____

____

____

For items 2 to 4, select the option that BEST completes the paragraphs.

2. Some nations won independence through a _______ or overthrow of the government. Jamaica gained its_______from the _________ of its______, Great Britain, in 1962.

   A. Nation, independence, colonial rule, revolution
   B. Revolution, nation, colonial rule, independence
   C. Independence, nation, colonial rule, revolution
   D. Revolution, independence, colonial rule, coloniser

Figure 26: *Her Late Majesty Queen Elizabeth II.*

3. Jamaica is no longer ruled by the______. It became an independent ______ with its own_______, a document that outlines rules of the government. However, Jamaica is still part of Great Britain's______. Rather than being ruled by the King or Queen of Great Britain, Jamaicans can choose their own leaders who can come from a _____. They have______or the right to vote.

   A. Monarchy, nation, constitution, commonwealth, political party, suffrage
   B. Suffrage, political party, monarchy, nation, constitution, commonwealth
   C. Commonwealth, suffrage, monarchy, political party, nation, constitution
   D. Monarchy, commonwealth, suffrage, constitution, nation, political party

4. With______, Jamaica was no longer ruled by another country. All Jamaicans were free to exercise their _______and vote for the political party they wanted. Workers benefited from ________that negotiated and protected their rights.

   A. Franchises, trade unions, self-government
   B. Self-government, trade unions, franchises
   C. Self-government, franchises, trade unions
   D. Trade unions, self-government, franchises

Objective: Use mathematical skills to construct timeline to show major developments in Jamaica's history from arrival of the Tainos to independence.

Item Type: **Single** selected response

Use the timeline shown to answer items 5 and 6.

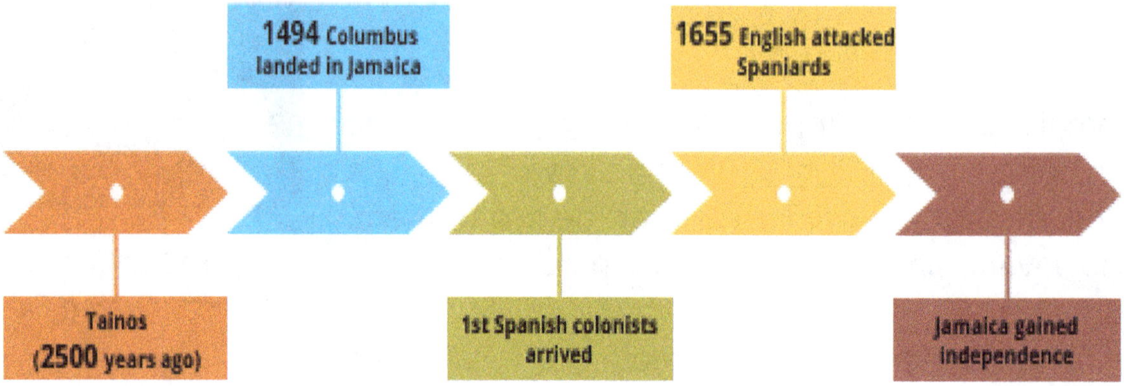

5A. In what year did Jamaica gain Independence?

A. 1655
B. 1962
C. 1956
D. 1494

5B. In what year did the Spanish colonists settle in Jamaica?

E. 1838
F. 1655
G. 1494
H. 1509

5. How long after Columbus landed in Jamaica did the English arrive?
   A. 161 years
   B. 1494 years
   C. 241 years
   D. 146 years

Objective: Examine, compare and evaluate multiple sources that outline the life and work of Marcus Garvey, Norman Manley and Alexander Bustamante.

Item Types: Single selected and short constructed responses

Read the passages below, then use them to answer items 7 and 8.

Source 1

Marcus Garvey was born in 1887 and recognised as Jamaica's first national hero in 1969. He believed in racial equality and black pride. He encouraged self-governance and self-reliance among blacks. In 1914, he founded the Universal Negro Improvement Association (UNIA) in Jamaica. Garvey then travelled to the United States and was imprisoned and deported in 1916 because he promoted freedom from oppression to blacks throughout the United States. He attempted unsuccessfully to form a political party in Jamaica a few years later. He died in 1940 in England.

Source 2

Norman Manley was born in 1893. A lawyer, he fought for the rights of workers and for improvements in their working conditions. He supported the trade union movement led by Alexander Bustamante. Manley strongly believed in democracy and in universal adult suffrage. He founded the People's National Party in 1938, helped to draft the Jamaican constitution and lobbied for Jamaica's independence. He died in 1969. He was honoured as a national hero after his death.

Source 3

Alexander Bustamante, cousin of Norman Manley, was born in 1884.

He made the public aware of the social and economic problems of Jamaica's poor. People became frustrated with unemployment and poor working conditions. This led to widespread riots and strikes. He formed the Bustamante Trade Union to protect the rights of the working class. He later founded the Jamaica Labour Party in 1943 and became Prime Minister of independent Jamaica in 1962.

6. Which source(s) SUPPORT(S) the statement below?

   National heroes helped Jamaica become a self-governing and an independent nation.

   A. Source 3 only
   B. Source 2 only
   C. Sources 2 and 3
   D. Sources 1, 2 and 3

8A. Which view do all three national heroes have in common?

   A. They thought that national pride was essential to Jamaica's freedom.
   B. They believed that the people would benefit from a self-governing and independent Jamaica.
   C. They opposed democratic ideas and values.
   D. They saw universal adult suffrage as the path to Jamaica's independence.

8B. How can you decide if the articles above are trustworthy? Identify THREE criteria that can help you to decide.

____

____

____

____

____

____

____

____

Objective: Apply lessons learnt from the lives of Marcus Garvey, Norman Manley and Alexander Bustamante to new situations.

Item Type: Short constructed response

Your parents have never voted because they believe it is a waste of time. They prefer to spend election day sleeping and, in the evening, they watch the results to see which party has won.

9. What have you learnt from Norman Manley that you could share with your parents to help them to see the importance of voting?

____

____

____

____

____

____

10. Your uncle has hired Mr Joel to clean the yard. Mr Joel is unemployed and says he will work for any amount. Use what you have learnt from Alexander Bustamante to encourage your uncle to pay Mr Joel a fair wage.

____

____

____

____

____

____

Marcus Garvey said,

'A people without the knowledge of their history, origin and culture is like a tree without roots.'

Do you agree? Can you imagine a tree having no roots? How would it grow? Do you know about your roots, your ancestors?

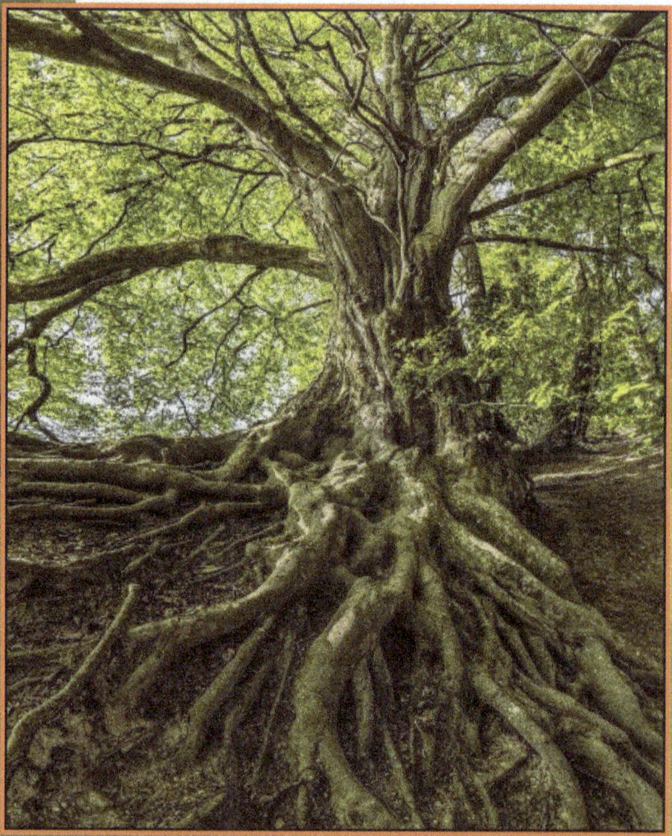

Objective: Compare the paths toIndependence taken by Jamaica, Haiti and Cuba.

Item Type: Single selected response

Read the statement below and use it to answer item 11.

The paths to Independence for Jamaica, Cuba and Haiti involved removing colony control either through force or diplomacy.

11. Which of these is the MOST LIKELY reason for each nation's revolt against colony control?

A. Taxation without representation
B. Poor treatment of the enslaved
C. Desire for each nation to be self-governing
D. Tough trade and labour policies

According to Thomas A. Edison, 'Our greatest weakness lies in giving up. The most certain way to succeed is always to try just one more time.'
Continue to answer the questions. Stay focused, do one question at a time and don't give up. Come wid mi. Mek wi continue.

Objective: Name the major personalities involved in the Independence movement in Jamaica, Cuba, Haiti.

Item Types: Order match and single selected responses

Read and answer items 12 and 13.

12. Below are three sets of national heroes involved in the Independence movement in their countries.

Box 1	Box 2	Box 3
José Martí Carlos Manuel de Céspedes y del Castillo	Marcus Garvey Norman Manley Alexander Bustamante	Toussaint L'Ouveture Jean Jacques Dessalines Henri Christophe

Choose the list that shows the countries from which each set of national heroes belongs List the countries from boxes 1 to 3.

A. Spain, Jamaica, France
B. Republic of Cuba, Jamaica, France
C. Spain, Jamaica, Republic of Haiti
D. Republic of Cuba, Jamaica, Republic of Haiti

13. Which leader was imprisoned in the United States of America, then deported for his views?

A. Fidel Castro
B. Norman Manley
C. Marcus Garvey
D. Jean Jacques Dessalines

Just for fun! Can you name these handsome Jamaican National Heroes?

 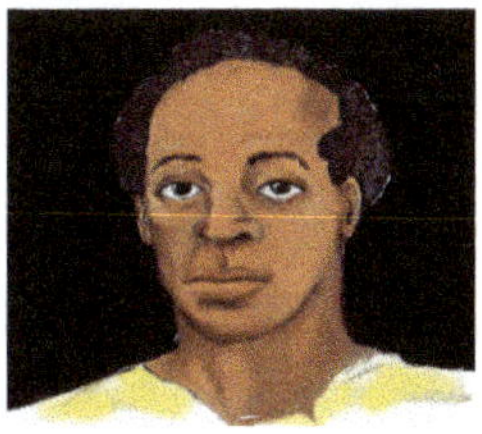 

________________ ________________ ________________

Objective: Compare how Independence is commemorated in Jamaica and other countries.

Item Types: Extended writing constructed and multiple selected responses

14. Write a letter to your friend in Canada telling him or her how Independence is commemorated in Jamaica and in one other Caribbean country. In the space below list the main points you would include in the body of the letter.

____

____

____

____

____

____

15. Which commemorative Independence activities do Haiti and Cuba share with Jamaica? Choose all that apply.

   A. food and drink
   B. parades
   C. dancing and music
   D. a day of silence

Objective: Explain the significance of Independence Day.

Item Type: Short constructed response

16. Independence Day is a national holiday in Jamaica. Why do you think this is so?

____

____

____

____

____

____

____

____

____

17. Identify TWO of Jamaica's Independence Day activities that show the importance of the day.

____

____

____

____

____

____

Objective: Formulate questions about Jamaica's decision to pursue Independenceand conduct research to answer these questions.

Item Type: Short constructed response

Mark is preparing for a debate. The moot is, 'Jamaica would have been better off had it not sought independence from England'. His team is supporting the moot.

18A. Write THREE questions that Mark's team would need to answer in order to win the debate.

____

____

____

____

____

____

18B. Write one possible answer to each of your questions.

____

____

____

____

____

____

____

____

Objective: Distinguish between dependent and independent countries in the Caribbean.

Item Types: Table grid and single selected responses

19. Some Caribbean countries are listed below. Use a tick (✓) to indicate if they are dependent or independent.'

Caribbean Nation	Dependent	Independent
Anguilla		
Barbados		
Jamaica		
The Bahamas		
British Virgin Islands		

20. Which of these is a characteristic of an independent Caribbean nation?

A. It has a participatory government.
B. Citizens do not have to pay taxes.
C. Citizens receive free healthcare, education and retirement.
D. They have a Governor General to represent the Prime Minister.

Objective: Show appreciation for the work done by individuals in Jamaica's decision to pursue Independence.

Item Type: Single selected response

21. Which action is MOST LIKELY to show appreciation for the work done by Jamaican nation builders?

A. The Jamaican flag is hoisted on Independence Day.

B. Restaurants serve some traditional Jamaican food.

C. The media create public awareness of the contributions they made towards Jamaica's independence.

D. Schools and businesses close to commemorate the persons who worked to make independence a reality.

Source: *https://jis.gov.jm/emancipendence-celebrations-important-minister-grange.*

Figure 27: *Patrons celebrate and wave flags at the Grand Gala Festival to celebrate Jamaica's independence and emancipation.*

Objective: Weigh the arguments for and against being an independent nation and draw conclusions about Jamaica's decision to pursue Independence.

Item Types: Single selected and short constructed responses

Work in pairs. Read the conversation between Patrice and her father.

22. One person will analyse the points from Patrice's perspective and the other will assess her father's perspective. You can summarise the main points about the pros and cons of Jamaica being independent in the table.

PROS	CONS

23. At the end of your discussion, answer this question: Was it a good decision for Jamaica to pursue independence?

____

____

____

Read the passage below and use it to answer items 24, 25 and 26.

After Jamaica's independence from Great Britain on August 6, 1962, the country became responsible for its own economy. Agriculture expanded and though sugar was the most exported crop, bananas ranked second and coffee and cocoa were also exported. Mining gradually became an important economic activity and Jamaica became the world's largest producer of bauxite and alumina.

Over time, tourism overtook agriculture, manufacturing and mining as the leading revenue source for the country. Jamaica remains a well-loved place to visit as tourists from all over the world come to experience the island's beauty, culture and people. Wages in Jamaica are still relatively low, but trade unions continue to negotiate for better wages and conditions for workers.

24. Which of the following BEST describes Jamaica's economy after Independence?
    A. Workers were paid more.
    B. There was more diversity in agricultural exports.
    C. There were more sources of revenue for the country.
    D. Trade unions negotiated for better working conditions.

25. According to the passage, what area of the economy is still a challenge?
    A. The damaging effects of mining on the environment
    B. Competition from other countries
    C. Low wages
    D. Bad associations with agriculture caused by the history of slavery

26. What conclusion can you draw about Jamaica's decision to pursue Independence?

   A. Jamaica was not ready to self-govern.
   B. It became a more resourceful and self-sufficient nation.
   C. Jamaican citizens had better opportunities for education and employment.
   D. Great Britain was wealthier and more powerful than Jamaica.

Source: *https://jis.gov.jm/pod/kamina-johnson-smith/*

Figure 28: *Minister of Foreign Affairs and Foreign Trade, Senator the Hon. Kamina Johnson Smith (left) and Chinese Ambassador to Jamaica, His Excellency Tian Qi (right).*

Objective: Resolve conflicts amicably while completing tasks in collaborative groups.

Item Type: Short constructed response

Read the scenario below, then respond to items 27 and 28.

Four students, Jermaine, Matthew, Camille and Francine, have been placed in a group in Social Studies class. They will remain in that group for the entire term. Francine is very careful and tries not to make mistakes. However, she works very slowly. Matthew does not like to listen or follow the instructions of others. Jermaine and Camille are easy going and want to finish work quickly so they can get on to important things like play time.

27. In your small groups, discuss TWO possible conflicts that may arise from the scenario given. For each conflict identified, state TWO ways it can be resolved.

____

____

____

____

____

____

____

____

____

____

28. Make a poster with your suggestions for resolving group conflicts that can be shared with your classmates.

Objective: Compile and arrange alphabetically a list of sources including author, title, publisher and date of publication.

Item Type: Short constructed response

Use the information below to answer items 29 to 30.

Junior has just completed a project on Jamaica's Independence. His teacher wants him to include a list of the sources he used, complete with the author's name, book title, publisher and publication date.

29A. Look at the reference list of books below and select the THREE most relevant ones you think Junior would have used. Help him to compile his reference list.

Title	Author	Publisher	Date of publication
A. Pre- independence History of Jamaica	**A.** Barbara Saddler	**A.** Beautiful Books	**A.** 1976
B. Slavery to Independence: A Caribbean Perspective	**B.** Garth Munroe	**B.** Begonia Publishers	**B.** 1965
C. Jamaica's Creative Past: Food, Festivals and Folklore	**C.** Gordon People	**C.** FSL Publishing Ltd.	**C.** 1995
D. Cricket, Colonialism and Caribbean Community	**D.** Collen Brooks	**D.** Carrington Caribbean	**D.** 2014
E. We Want Justice! Jamaica's Journey to Nationhood	**E.** Hope Dennis	**E.** Jamdown Publishers	**E.** 2009

30. Find FIVE additional sources. List your sources alphabetically and with all the relevant information.

STOP

EVALUATE YOUR WORK!

Remember to evaluate your work by completing the table at the beginning of the unit. You can always review the concepts as well as ask for help.

I understand.	**I somewhat understand.**	**I don't understand.**

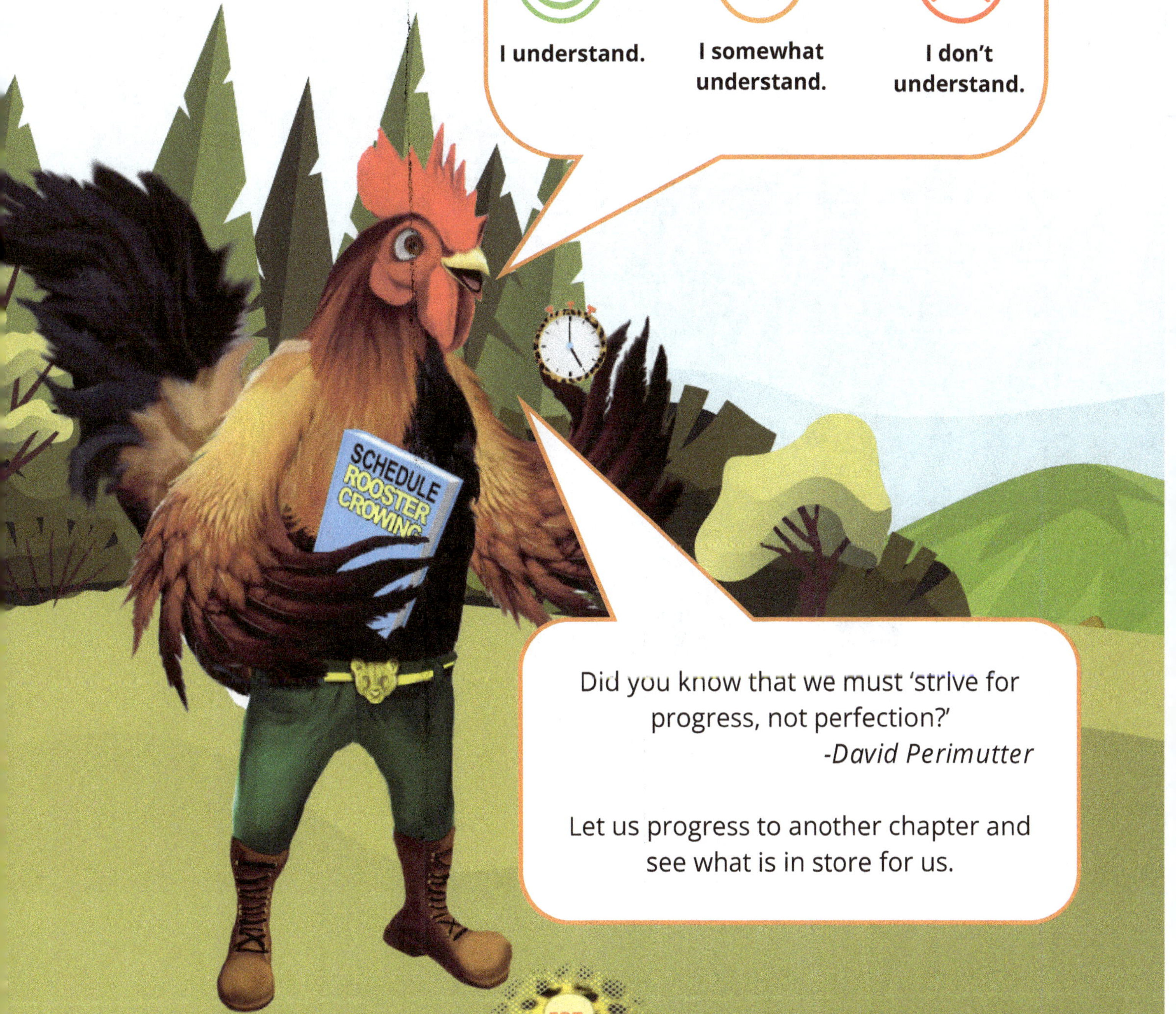

Did you know that we must 'strive for progress, not perfection?'

-David Perimutter

Let us progress to another chapter and see what is in store for us.

Unit Assessment Based on the NSC

Term 1, Unit 3

Focus question: How do we show respect and loyalty to our country?

'Loyalty to the family must be merged into loyalty to the community, loyalty to the community into loyalty to the nation, and loyalty to the nation into loyalty to mankind. The citizen of the future must be a citizen of the world.'

– Thomas Cochrane

TERM 1, UNIT 3

Living Together

I need to be able to:	Column A I completed the following:	Column B How did I do?		
		1. I understand.	2. I somewhat understand.	3. I don't understand.
Develop working definitions and use correctly the following concepts: emblem, flag, coat of arms, symbols, nationhood, anthem, crest, bearing, motto, patriotism				
Explain what national symbols and emblems are and analyse their importance to nationhood.				
Identify and describe the national symbols of Jamaica.				
Examine images which show the changes in the Jamaican coat of arms and justify the changes made.				
Tell the meaning of each colour on the Jamaican flag.				
Recite and interpret the national anthem and national pledge.				

I need to be able to:	Column A I completed the following:	Column B How did I do?		
		1. I understand.	2. I somewhat understand.	3. I don't understand.
Select a group/organisation, describe its purpose and values, then design symbols to reflect its purpose and values.				
Observe etiquette to be adhered to regarding national symbols and emblems.				
Assess the role and responsibilities of the citizens in acknowledging and maintaining pride in our national symbols.				
Show respect for our national symbols and emblems.				
Show respect for individuals who hold public offices.				
Negotiate and compromise to resolve conflict during collaborative work.				

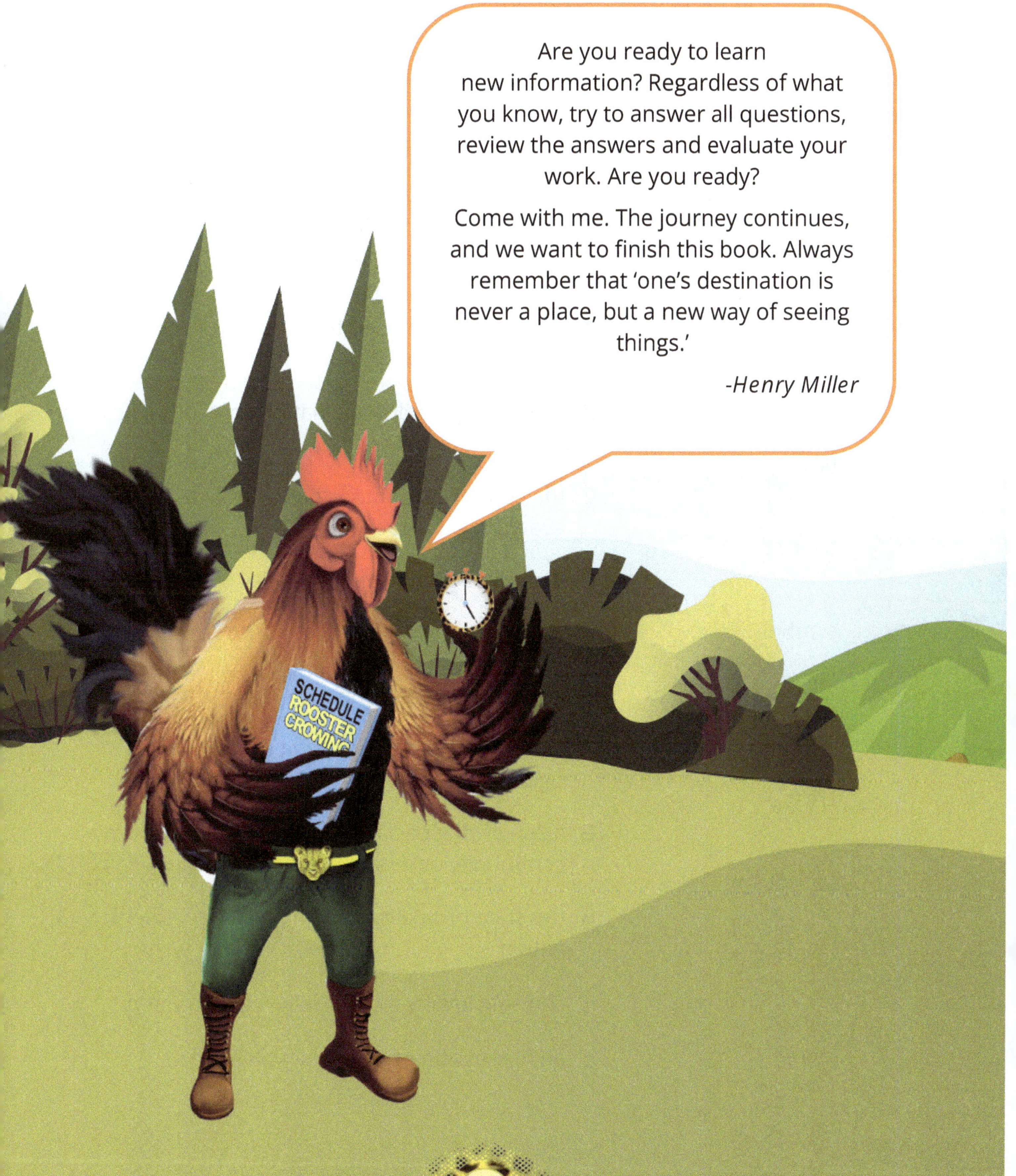
Are you ready to learn new information? Regardless of what you know, try to answer all questions, review the answers and evaluate your work. Are you ready?
Come with me. The journey continues, and we want to finish this book. Always remember that 'one's destination is never a place, but a new way of seeing things.'
-Henry Miller
SCHEDULE ROOSTER CROWING

Objective: Develop working definitions and correctly use the concepts: emblem, flag, coat of arms, symbols, nationhood, anthem, crest, bearing, motto, patriotism.

Item Types: Order match, single selected and short constructed responses

Column B has definitions that should be used for both items 1A and 1B.

1A. The table below contains a list of words in column A and matching definitions in column B. In column A, write the letter for the corresponding definitions on the lines given. Each word should be used only once.

Column A	Column B
	A. anything that is used to stand for or represent something else
____ emblem	B. a design symbolising a nation, institution or family
____ flag	C. the central part of the coat of arms to which everything is attached
____ bearing	D. a piece of cloth having specific colours and patterns used as a symbol representing a nation
____ symbol	E. strong support of one's country
____ nationhood	F. an object used to represent something of significance
____ crest	G. a song with special meaning for people from a community or nation
____ patriotism	H. the state or quality of being an independent nation
	I. showing loyalty to your fellow countrymen
	J. an expression of beliefs or values

1B. Write definitions for the words below:

A. motto

____

____

B. anthem

____

____

C. coat of arms

____

____

Read the passage and answer the questions.

emblem, flag, reverence, coat of arms, symbol, nationhood, anthem, crest, bearing, motto, patriotism

Obtaining ______________ or national independence is a significant achievement. To represent and commemorate its newfound freedom, a nation will put special emphasis on the ______________(s), which represent different aspects of its country. For example, the country may have an ______________, which is a song of praise and respect. A country may also have a ___________ to express its beliefs or values.

Another common national emblem is a ________________ which students or military representatives may raise and salute at official events. Additionally, there is generally a rise in ________________because people's support of and pride in their own country increases after Independence.

Jamaica has all of these emblems to commemorate its independence. Jamaicans take pride in our__________________ 'Out of Many, One People' and our ________________with its special shield, crest and arms.

2. Which of the following lists the correct sequence of words to complete the passage above?
   A. emblem, flag, reverence, coat of arms, symbol, nationhood, anthem, crest
   B. nationhood, emblem, anthem, crest, flag, patriotism, motto, coat of arms
   C. emblem, reverence, flag, symbol, patriotism, crest motto, bearing, coat of arms
   D. nationhood, emblem, anthem, motto flag, patriotism, crest, coat of arms

'Recognize yourself in he and she who are not like you and me.'

- *Carlos Fuentes*

Objective: Explain what national symbols and emblems are and analyse their importance to nationhood.

Item Types: Single selected and short constructed responses

3A. Which group of special symbols and emblems would a newly independent nation want to focus on?

A. Country name, flag and constitution
B. Flag, flower and peace symbol
C. Constitution, country name and anthem
D. Flag, anthem and flower

3B. Nadia is having an issue differentiating between the terms 'national symbols' and 'national emblems'. Use examples to explain the difference to Nadia.

____

____

____

____

____

____

____

____

____

____

____

____

3C. If you had a chance to create another national symbol for Jamaica, what would it be? Give a reason for your choice.

3D. Nadia does not think having national symbols matter. What would you say to convince her that they are important for an independent nation?

Objective: Identify and describe the national symbols of Jamaica.

Item Types: Single selected and short constructed responses

4. Which of the following are the national symbols of Jamaica?

   A. National flag, coat of arms, anthem
   B. National flag, Blue Mahoe, Lignum Vitae, ackee
   C. Ackee, humming bird, Blue Mahoe, Lignum Vitae
   D. Ackee, Lignum Vitae, coat of arms, Blue Mahoe

5. Which of the above images represents the Jamaican national bird?

A.

B.

C.

D.

A. bird A   B. bird B   C. bird C   D. bird D

Sources: *Wikipedia and DigiJamaica.com*

6A. Write at least THREE things you know about the Blue Mahoe.

____

____

____

6B. Describe the national flower in terms of its shape, colour and size.

____

____

Objective: Examine images which show the changes in the Jamaican coat of arms and justify the changes made.

Item Types: Single selected and short constructed responses

Examine the images of the Jamaican coat of arms below and use them to answer items 7 and 8.

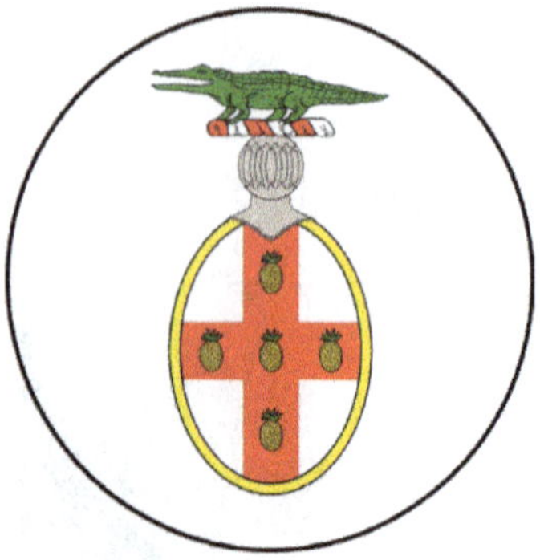

A. 1692

B. 1959

C. 1962

7. Which of the following options is one of the changes made to the motto after 1959?

   A. A cross was added to the centre of the coat of arms.
   B. Indigenous people were added to the coat of arms.
   C. Jamaica's name was added to the top of the coat of arms.
   D. The motto was written in English.

8. Why do you think changes were made to the Jamaican coat of arms over time?

____

____

____

Objective: Tell the meaning of each colour on the Jamaican flag.

Item Type: Sngle selected response

Item 9 refers to the Jamaican national flag.

Source: *https://jis.gov.jm/information/symbols/jamaica-national-flag/*

9. The Jamaican flag contains the colours green, black and gold. What do the black and gold represent?

   A. Creativity and strength and richness and sunshine
   B. Strength, sunlight's beauty and wealth of the country
   C. Hope and wealth of the country
   D. Hope and agricultural resources

I have something for you to think about. 'You've got to get up every morning with determination if you're going to go to bed with satisfaction.'

These words were spoken by George Lorimer.

Are you determined? Let me help you to prepare so you will be satisfied that you did your best to prepare for your social studies PEP exam.

Objective: Recite and interpret the national anthem and national pledge.

Item Type: Single selected response

Recall and say the words of the Jamaican national anthem and then answer item 10.

10. Which of these is the **MOST LIKELY** meaning of the words 'keep us free from evil powers?'

    A. The years of colonial rule that exposed us to evil In Jamaica
    B. Protect Jamaicans from those who may want to harm us.
    C. Evil powers used slave labour and created a harsh life for the enslaved..
    D. Evil powers could be Jamaica's own leaders and citizens.

Recall and say the words of the Jamaican pledge and use them to answer item 11.

11. What is the best conclusion that can be made about each citizen's responsibility to Jamaica?

    A. The Jamaican people must pledge their loyalty to the monarch of England.
    B. Each citizen in Jamaica is responsible for promoting the positive growth of Jamaica.
    C. The government will help the citizens to meet their responsibilities.
    D. Jamaicans are responsible for the well-being of Jamaicans around the world.

Objective: Select a group/organisation; describe its purpose and values, then design symbols to reflect its purpose and values.

Item Types: Single selected and short constructed responses

Read the following passage and use it to answer items 12 and 13.

Jamaica partners with Food for the Poor (FFP) to provide direct relief to poor people with urgent needs. Their goal is to help men, women and children in need of housing, food, medicine, water and education. They also aim to instill confidence and hope in the people they serve. They value helping all people without discriminating based on race, status or religion. They believe their work supports their mission to serve Christ by helping those with the greatest needs.

12. Based on the passage about the Food for the Poor organisation, draw a symbol that could represent their purpose and values.

13. Which of these statements BEST describes the impact Food for the Poor has on the people of Jamaica?

   A. They improve the health and nutritional needs of poor people.
   B. They provide valuable jobs to boost/help the economy.
   C. They are an emergency relief organisation.
   D. They work to ensure that men, women and children receive medical, financial, social or spiritual help.

14. You are forming an environmental club at school. You and your friends agree that it is important to have a statement that describes the purpose and values of the club. Write your statement below.

____

____

____

____

____

____

____

____

____

Objective: Observe etiquette to adhere to regarding national symbols and emblems.

Item Type: Short constructed response

Read the statement and respond to items 15 A to C.

National symbols and emblems are important parts of a nation's history, culture and traditions and have different etiquette requirements.

15A. Identify ONE way to show respect for Jamaica's national flag at an official event.

____

____

____

____

15B. How should the Jamaican flag be positioned in the presence of other flags flown in Jamaica?

____

____

____

____

15C. What are persons expected to do as the Jamaican national anthem is being played or sung?

____

____

____

Objective: Assess the roles and responsibilities of the citizens in acknowledging and maintaining pride in our national symbols.

Item Type: Short constructed response

16A. Choose TWO Jamaican national symbols and discuss how citizens contributed to the selection of these symbols.

16B. Describe TWO of the main responsibilities that Jamaican citizens have in maintaining pride in their national symbols.

Objective: Show respect for our national symbols and emblems.

Item Types: Short constructed response and single selected responses

17. You have a new pen-friend who lives in Singapore. Using examples, tell your friend TWO ways Jamaicans can show that they respect the national symbols and emblems.

____

____

____

____

____

____

Use the following statement to answer item 18.

> The colours were so faded that they no longer represented Jamaica and could no longer be used in ceremonies. It had to be destroyed.

18. What national Jamaican symbol or emblem is being respected by this action?

A. The Jamaican coat of arms
B. The Jamaican flag
C. The Jamaican constitution
D. The national anthem

'Always remember you have within you the strength, the patience, and the passion to reach for the stars, to change the world.'
- Harriet Tubman
Come wid mi.
Mek wi prep for PEP.

Objective: Show respect for individuals who hold public office.

Item Type: Single selected response

Use the following statement to answer items 19 and 20.

Observing official protocols shows respect for those in public office. It shows you appreciate the significance of the office and the individual's service to Jamaica.

19. Which of these should be used as a verbal reference to the Governor General?

A. Greetings Governor General
B. Greetings Sir
C. Your Excellency
D. Your Governor General

20. Which is the proper protocol to announce the Prime Minister's arrival, if the Governor General is not in attendance at an official function?

A. Play the national anthem.
B. Practise a moment of silence.
C. Display the Jamaican flag in unison with all the other flags.
D. Everyone should be quietly seated with hands in laps.

Objective: Negotiate and compromise to resolve conflict during collaborative work.

Item Type: Short constructed response

Use the scenario below to respond to items 21A and 21B.

Johnny and Paula are excellent students. Whenever they get assignments, they work hard to get maximum scores. They do not like to work in groups because this usually takes longer and some group members do not do what they are supposed to do. These two students were paired to complete a project. Johnny wanted to look at the contribution of nation builders to our history and Paula thought it would be better to look at current leaders. They could not agree and neither student would compromise. Both students did two separate projects. However, the teacher insisted that they work together to submit one project.

21A. What would you say to Johnny and Paula to help them see the need to work together?

____

____

____

____

____

____

21B. Put yourself in the teacher's position. How would you help Johnny and Paula reach a compromise so that they could work together.

____

____

____

____

STOP

EVALUATE YOUR WORK!

Remember to evaluate your work by completing the table at the beginning of the unit. You can always review the concepts as well as ask for help.

I understand.	I somewhat understand.	I don't understand.

How are you doing?
Are you having fun? Have you learnt anything new? Stop and take the time to teach someone at home.
Three more chapters to go!
I don't know about you, but I study for long hours to get the grades I want.

Come wid mi. The journey continues.

Unit Assessment Based on the NSC

Term 2, Unit 1

Focus question: How do we show respect and loyalty to our country?

'Today is your day! Your mountain is waiting, so... get on your way.'

– Dr. Seuss

TERM 2, UNIT 1

The Physical Environment and Its Impact on Human Activities

I need to be able to:	Column A I completed the following:	Column B How did I do?		
		1. I understand.	2. I somewhat understand.	3. I don't understand.
Develop working definitions for, and correctly use the following concepts/terms: • mountain • mountain range • hill • valley • plateau • landforms • plains • forest reserve • summit • slopes.				
Create a thematic map showing the name and location of the major mountains in Jamaica.				
Use different criteria to rank mountains and mountain ranges.				
Use data to make comparisons and draw conclusions about how mountains affect weather and climate.				

I need to be able to:	Column A I completed the following:	Column B How did I do?		
		1. I understand.	2. I somewhat understand.	3. I don't understand.
Gather information from multiple sources and use it to describe the activities, goods produced and services that are carried out/ offered in mountain/hill environments, then draw conclusions about the importance of mountain/hill environments.				
Gather information from multiple sources and use it to analyse the effects of human activities on mountains.				
Design models and develop strategies to reflect best practices for human activities in mountain/hill environments.				
Make decisions that show responsibility and care for the environment.				
Critique the work/ideas of group members.				

STOP

REMINDER: Review the key concepts section before you continue.

Objective: Develop working definitions for and use correctly the following concepts/terms: mountain, mountain range, hill, valley, plateaus, landforms, plains, forest reserve, summit, slopes.

Item Types: Single selected, order match and table grid responses

Use the following words to fill in the blank spaces in the paragraph below. Some of the words may be used more than once.

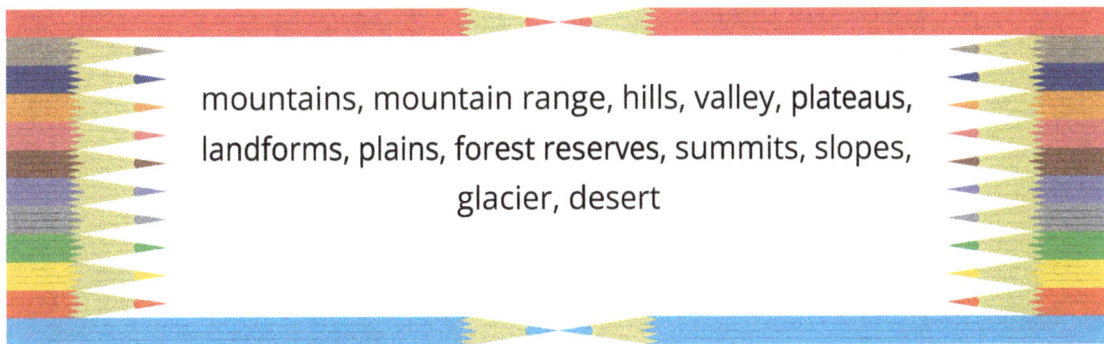

mountains, mountain range, hills, valley, plateaus, landforms, plains, forest reserves, summits, slopes, glacier, desert

1A. Jamaica is made of many different types of ________________, which are the natural geographic features of the Earth's surface. ________________ are ________________ that rise above the land that surrounds them. They form peaks, which are known as ________________. Several ________________ in a line together form a ________________. ________________ are also elevated lands but are smaller than ________________. They are generally rounded at the top.

Both ________________ and ________________ have ________________, which are the sides of these landforms. The slopes of the mountain or hill form a ________________, which is a low-lying area, generally in the form of a U or V shape that often has a river running through it. ________________ are higher lands that form a flat elevated plain. ________________ are lower flat areas, often good for farming. Throughout ________________ Jamaica, we have several ________________ ________________; forests that have many trees that are protected from being cut down.

1B. The table below lists some physical features found around the world. Use a tick (✓) to indicate those that are found in Jamaica.

Physical Feature	Tick (√) if found in Jamaica
desert	
valley	
glacier	
coastal plain	
mountain range	
tundra	
hill	

Are there any words on this page that you did not know? Take the time to learn and apply new words; use them in your daily conversations.
Do you agree with this quote by Henry Hazilt? 'A man with a scant vocabulary will almost certainly be a weak thinker... Knowledge of things and knowledge of the words for them go together. If you do not know the words, you can hardly know the thing.'

1C. Write the letter for the corresponding concept on the lines given in the table.

Concept	Definition
___ mountain	A. a landform that rises slightly higher than the area around it and often has a round top
___ mountain range	B. the sides of mountains or hills
___ hill	C. the natural geographic features on Earth's land surface
___ valley	D. a series of mountains that rise above the surrounding area and are usually arranged in a line,
___ plateau	E. the top of a mountain or hill
___ landform	F. a high, flat area that rises above the surrounding region
___ plains	G. a forested area that is protected by law
___ forest reserve	H. a vast region of flat or gently sloping land, often used for farming
___ summit	I. a high landform, usually with steep sides, that rises 300 metres or more above the surrounding area
___ slope	J. low-lying, 'U' or 'V' shaped region usually found between hills or mountains. Rivers or streams usually flow through it.

2. The Pedro Plains in St. Elizabeth were created by large deposits of sediment flowing down from the mountainous areas. Which of the following features do plains offer?

A. Rocks that are good for stone buildings
B. Sandy valleys rich in minerals for mining
C. A relatively flat piece of land which can be used for farming
D. High mountains good for the tourist industry

Which of the following lists the correct sequence of words to complete the passage below?

3. The island of Jamaica is made up of many different types of ________________. The lowest in elevation are areas along the coast known as ________________. These areas have rich, fertile soil, making them great for farming. The highest elevation can be found in the ____________________. For example, Blue Mountain Peak rises 2,256 metres above sea level. At the very top or the highest point of the mountain, you find the ____________________. Between the mountain ranges, the slopes of the mountains form U shaped or V shaped areas known as ____________________. These areas often have rivers running through them.

   A. Mountain ranges, valleys, summit, plains, landforms
   B. Summit, landforms, valleys, plains, mountain ranges
   C. Plains, summit, mountain ranges, valleys, landforms
   D. Landforms, plains, mountain ranges, summit, valleys

Which of the following lists the correct sequence of words to complete the passage below?

4. Landforms can be differentiated by their height above sea level. The highest type of landform is a ________________. Another elevated landform that is often smaller than a mountain is known as a ______________. It has a rounded top and is not as high as a mountain. Both mountains and hills have sides that are called ____________________. They form the valleys. Another type of flat elevated area is known as a ________________. Throughout all these landforms, we find protected areas of trees known as ____________________ ______________.

   A. Mountain, hill, slopes, plateau, forest reserves
   B. Forest reserves, mountain, slopes, hill, plateau
   C. Slopes, hill, mountain, forest reserves, plateau
   D. Plateau, forest reserves, mountain, hill, slopes

Objective: Create a thematic map showing the names and locations of the major mountains in Jamaica.

Item Type: Single selected response

Examine the physical map of Jamaica and use it to answer item 5.

Source: *https://commons.wikimedia.org/wiki/File:Jamaica_relief_location_map.jpg.*

5A. Use the letters A to E to label the approximate location of the mountains listed on the map of Jamaica.

A. Mocho Mountains
B. Blue Mountains
C. Don Figuerero Mountains
D. Dry Harbour Mountains
E. John Crow Mountains

5B. Based on the map, we can conclude that Jamaica's Blue Mountains range is to the _________ section of the island.

A. Western
B. Northern
C. Central
D. Eastern

SCHEDULE
ROOSTER
CROWING
Did you know that
in real life we use the word mountain
and related words such as hills and
long roads to describe problems?
What kind of mountains are you
currently facing? Can you
deal with your mountains by yourself?
Where do you go for help?
Come wid mi. Let's move some of
these mountainous questions.

Objective: Use different criteria to rank mountains and mountain ranges.

Item Types: Table grid, order match and single selected responses

The chart shows some of the mountains found on the island of Jamaica.

6A. Rank the mountains 1 to 5 from LOWEST to HIGHEST above sea level.

Name of Mountain	Parish	Rank (1=lowest 5=highest)
Gossomer Peak	Saint Thomas	
Dolphin Head	Hanover	
Blue Mountain Peak	Portland/Saint Thomas	
High Peak	Portland	
Frasers Mountain	Hanover	

6B. Jamaica's mountains are centrally located, with its ranges running west to east. Order the mountains and mountain ranges below based on their location on the island.

Don Figuerero Mountains

Orange Hill

Blue Mountains

Dolphin Head

Mocho Mountains

John Crow Mountains

Dry Harbour Mountains

Santa Cruz Mountains

Western	Central	Eastern

Examine the table below and use it to answer item 7.

Mountain Range	Highest point of elevation in metres
Blue Mountains	2,256
Dry Harbour Mountains	762
John Crow Mountains	1,140
Mocho Mountains	499

Source: *https://www.britannica.com/place/Jamaica.*

7. Based on the information from the chart, which statement is CORRECT?

   A. The Mocho Mountains are the highest on the island.
   B. The John Crow Mountains are lower than the Dry Harbour Mountain range.
   C. The Blue Mountain Peak is over four times higher than the Mocho Mountains' highest point.
   D. The Dry Harbour Mountains are the lowest on the island.

Source: *https://commons.wikimedia.org/w/index.php?curid=24023468.*

Objective: Use data to make comparisons and draw conclusions about how mountains affect weather and climate.

Item Types: Single selected and short constructed responses

Use the diagram below to answer item 8.

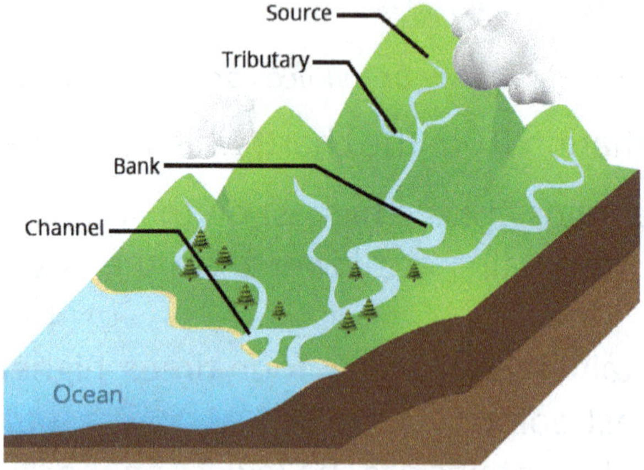

8. On which side of the mountain range would you expect to find more rainfall?

   A. On both sides
   B. On the windward side
   C. On the leeward side
   D. At the top of the mountain range

9. Shanique was on vacation with her family. On Monday, they went to Winnifred Beach and the next day they travelled to Blue Mountain Peak. She was surprised that even though it was so hot at the beach, she needed a sweater on the Blue Mountains. Explain why there was such a difference in the temperature between the two locations.

____

____

____

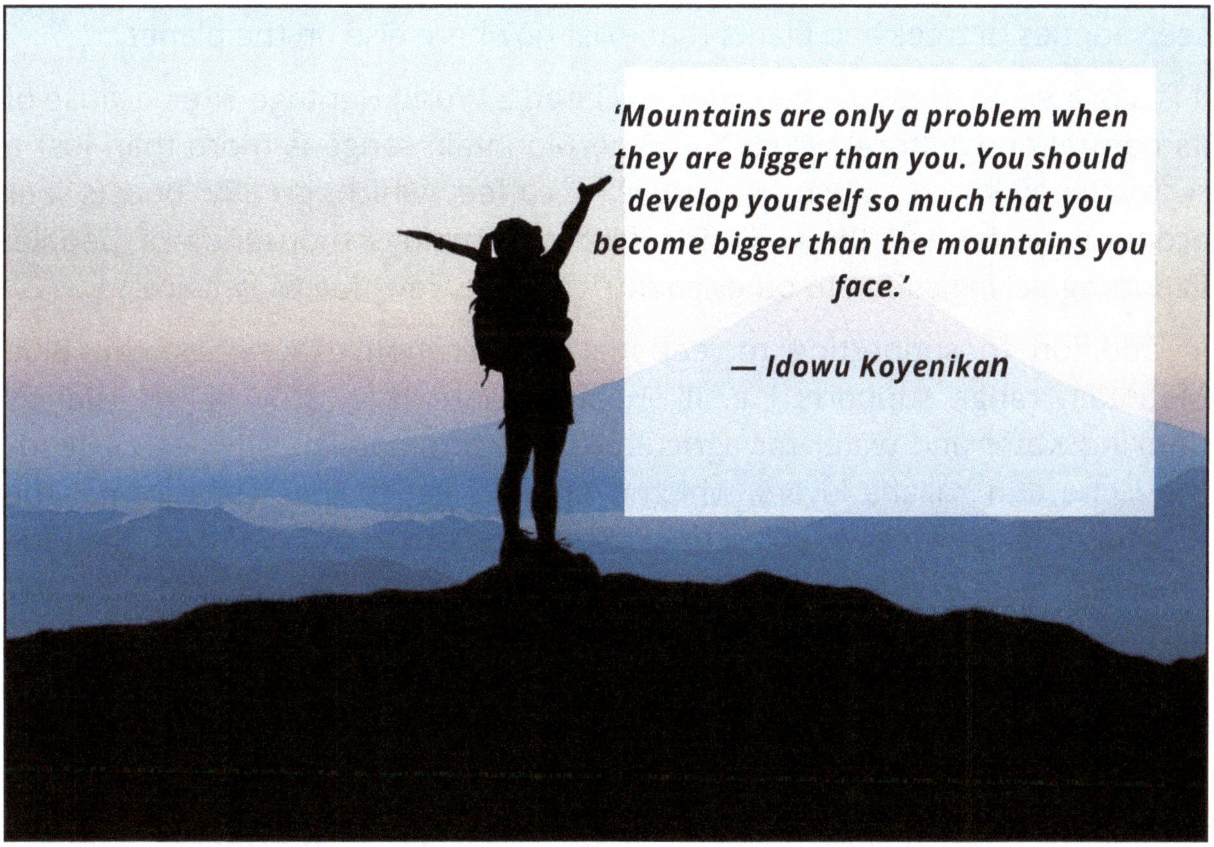
'Mountains are only a problem when they are bigger than you. You should develop yourself so much that you become bigger than the mountains you face.'
— Idowu Koyenikan

Objective: Gather information from multiple sources and use it to describe the activities, goods produced and services that are carried out/offered in mountain/hill environments, then draw conclusions about the importance of mountain/hill environments.

Item Type: Extended writing constructed response

Use the following excerpt to respond to item 10.

'Why Protect our Mountains?'

Jamaica's Blue Mountain range is sought after for its beauty and tranquility. Residents and tourists alike visit the mountains to hike, camp, bird watch and see varieties of trees and plants that exist nowhere else on the planet.

It is a protected area that has been declared a World Heritage Site because of its natural and cultural value. The Blue Mountain range is more than just a recreational space. Much of Jamaica's coffee, which greatly boosts our economy, is farmed there. Coffee farming employs thousands of people. Exporting/selling coffee to other countries brings revenue to Jamaica.

In addition to supporting recreational and economic activities, the Blue Mountain range supports life. It supplies more than 30% of the island's drinking water and water for agriculture and commercial activities. It feeds the Hope and Yallahs Rivers, which eventually get to the Mona Dam. The Wagwater River, which supplies the Hermitage Dam, is also fed from the Blue Mountain range. Altogether, the Blue Mountains, like other hills and mountains in Jamaica, are important and valuable.

10. Write an essay using the source above and 2 others sources of information that show the importance and value of mountains and hills to Jamaica's environment. Highlight and use information from all THREE sources to write the essay titled 'Why Protect our Mountains?

____

____

Objective: Gather information from multiple sources and use it to analyse the effects of human activities on mountains.

Item Types: Short constructed and extended writing constructed response

Respond to the following items. Use multiple sources of information, for example, interviews, articles or informative videos to help you.

11A. Write TWO types of human activities that have harmful effects on Jamaican mountains.

____

____

11B. Explain ONE long-term and ONE short-term effect of each of these activities.

____

____

____

____

____

____

____

____

11C. What are some possible solutions to the long and short term harmful effects of human activity on Jamaican mountains?

____

____

Objective: Design models and develop strategies to reflect best practices for human activities in mountain/hill environments.

Item Types: Short constructed and extended writing constructed response

Read the statement and answer item 12.

Jamaica has 114 forest reserves plus several national parks. We need these areas to protect biodiversity.

12. Design a model using circles and boxes. Make the title of your model be 'Protecting Forest Reserves and National Parks'. In your model, list some strategies, policies and actions individuals can use to take care of these environments.

Read the sentence below and use it to answer item 13.

Some harmful species that are NOT native to Jamaica, like the white ginger lily and the vampire fern, are destroying the indigenous plants of many hillsides across Jamaica.

13. Identify TWO strategies that the agriculture ministry could use to reduce the population of white ginger lily and vampire fern.

Objective: Create thematic map showing the names and locations of major mountain ranges of the world (at least one in each continent).

Item Type: Single selected response

Examine the map of the world below and use it to answer item 14.

Source: *clip-art-blank-world-map-world-map-outline.png*

14. The mountain ranges below have been assigned the numbers 1 to 5. Insert the numbers on the world map to show their approximate location on the continents.

    1. Andes
    2. Himalayas
    3. Rockies
    4. Great Dividing Ranges
    5. Alps

15. On which continent would you find Mount Everest, the highest mountain in the world?

    A. North America
    B. Asia
    C. Africa
    D. South America

SCHEDULE ROOSTER CROWING
We will review caring for the environment next.
Do we
recycle bottles in Jamaica? How can you help or start such a programme in your community? Do you think you are
too young? Nonsense!
Read about a young Swedish climate activist, Greta Thunberg, who won a coveted place as the youngest person to ever be on the page of a very popular magazine, the Time Magazine.
If she can be influential, so can you.
Come wid mi.

Objective: Make decisions that show responsibility and care for the environment.

Item Types: Single selected and multiple selected responses

Read the paragraph below and use it to answer item 16.

Ayana and Rashona had just finished eating their lunch by the beach. Rashona's waste from lunch had included a plastic water bottle that she had filled with lemonade. Rashona wanted to throw the empty bottle in the regular garbage bin.

16. What advice can Ayana give to Rashona to help her show greater care for the environment?

A. Throw the bottle on the beach so the sea animals can swallow it.
B. Recycle the water bottle to prevent it from ending up in the trash.
C. Bury the water bottle in the sand.
D. Put the water bottle in the ocean where it can sink to the bottom.

Examine the bulleted list and image below, then answer item 17.

The Jamaican Black Bird

- Found only in Jamaica
- Its habitats are the Cockpit Country and some sections of the Blue and John Crow Mountains
- Threatened by bauxite mining, deforestation, farming and burning of charcoal

Source: *https://en.wikipedia.org/wiki/Jamaican_blackbird.*

17. What are TWO ways to keep our native black bird safe?

A. Pass a law to protect its habitat
B. Support deforestation
C. Start a conservation group that educates the surrounding communities
D. Introduce more predators to the area

Read the statement below and use it to answer item 18.

A coffee farmer notices that he keeps losing soil from his mountainside farm each year after the rainy season.

18. What would be the BEST solution for the farmer to prevent further erosion?

    A. Remove more trees to make room for more farming.
    B. Build a road to remove the soil before erosion occurs.
    C. Do terracing and contour farming, plant deep rooted trees and cover crops
    D. Dig a trench where the soil can flow.

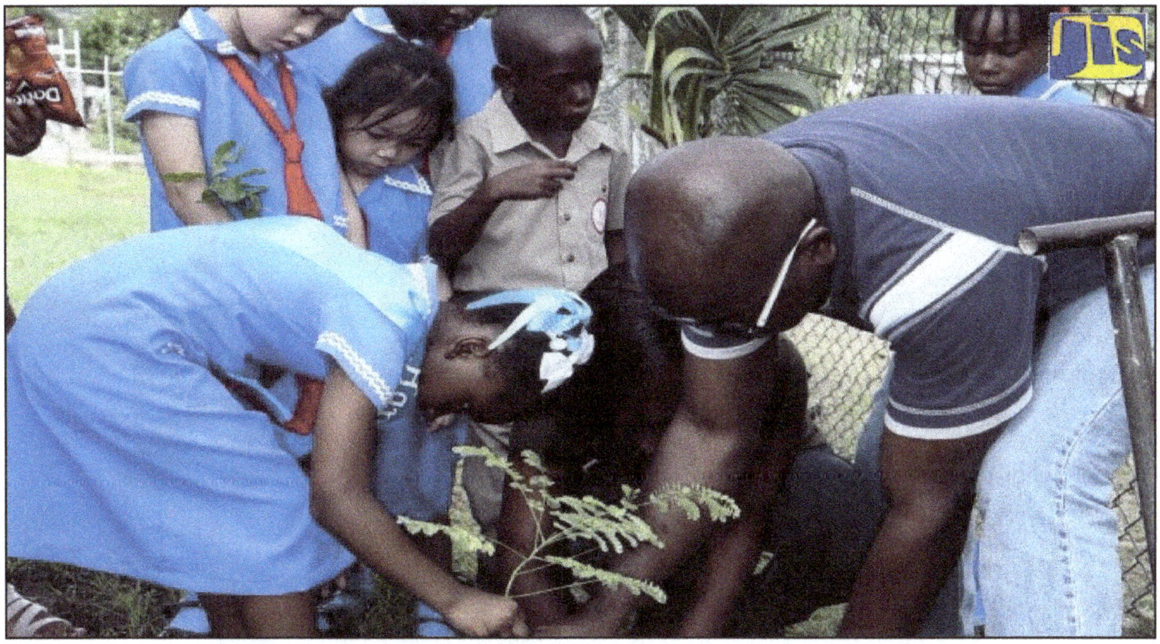

Source: *https://jis.gov.jm/photos-senator-the-hon-pearnel-charles-jr-at-unesco-tree-planting-initiative- trinity-preparatory-school.*

Figure 29*: Planting trees helps prevent soil erosion. Minister without Portfolio in the Ministry of Economic Growth and Job Creation, Senator the Hon. Pearnel Charles Jr. (right) and student of Trinity Preparatory School in Linstead, St. Catherine, Jhariel Davis (left), plant a sapling during UNESCO Youth Advisory Committee tree planting and renovation initiative at the institution. Looking on are other students of the school. Photo by Rudranath Fraser.*

Objective: Critique the work/idea of group members.

Item Types: Table grid and short constructed responses

Read the following passage about deforestation and use it to answer item 19.

There is a genuine need for humans to remove trees. We need wood for housing, paper and fuel. We also need space for housing and farming. However, deforestation or the removal of forests has increased to dangerous levels. As a result, we face soil erosion, poor air quality, increasing global warming, loss of habitats for forest animals and declining rainfall.

Soil erosion removes nutrients from the soil. It leads to landslides, mudslides and flash floods. Removing forests also reduces air quality. Trees provide oxygen and remove carbon dioxide, dust and other pollutants from the atmosphere. Increased carbon dioxide in the atmosphere is one of the major contributors to global warming.

Trees help to regulate temperature. They provide homes and food for our animals. The Jamaican iguana, for example, is facing extinction because of deforestation. Finally, deforestation results in less rainfall. Longer and more severe droughts will not only affect how much water we have, but how much food we have as well. There is a genuine and urgent need to protect and restore our forests.

19. Use a tick (✓) to indicate whether each argument is supported or not supported by the passage.

Argument	Supported	Not Supported
Soil erosion is the main effect of deforestation.		
Most forests have remained untouched.		
The cutting down of forests affects the environment in many ways.		
Deforestation also affects Jamaica.		

20. You are working in a group to write a report on some of the effects of deforestation. Ayana is the Research Manager in your group. She presented the information below on the effects of deforestation at the class group meeting.

A. Soil erosion
B. Droughts
C. Road to remove the soil before erosion occurs.
D. Loss of biodiversity
E. Rock wall or retaining wall
F. Climate change

Was the information completely accurate? Write a paragraph explaining your views.

____

____

____

____

____

____

I have a special assignment for you.
Work with your peer (another student) to review each other's responses to question 20.

What exactly will you be doing? You will review and critique how the other person responded to the questions, and then provide feedback. Here are some suggestions for giving useful feedback:

1. Spend a few minutes to review all the answers.
2. Explain your partner's strengths and weaknesses. What did the person do well and what could be improved?
3. Offer any suggestions, for example, study tips or concepts, that your peer may be missing.
4. Be as clear and specific as possible.
5. Be courteous and kind; don't be rude with your feedback.
6. Refer to any additional resources, such as your textbook, for any concept that is unclear to you both.
7. Have fun and learn from each other.
8. Evaluate each other on the next page.

EVALUATE YOUR WORK!

Remember to evaluate your work by completing the table at the beginning of the unit. You can always review the concepts as well as ask for help.

I understand.	**I somewhat understand.**	**I don't understand.**

Hannah Hudson believes, 'Everything you don't know is something you can learn.'

What do you believe? Are you ready to learn some more? You can continue to work on your own or continue to work with your partner. Do what works best for you while you are studying but remember that you work alone on test day: 'Every tub haffi siddung pon dem own bottom,' as the old folks used to say.

We are nearing the finishing line.

Come wid mi.

Unit Assessment Based on the NSC

Term 2, Unit 2

Focus question: 'How can we classify the landmasses and water bodies of the world?'

'When the last tree has been cut down, the last fish caught, the last river poisoned, only then will we realize that one cannot eat money.'

– American Indian proverb

TERM 2, UNIT 2

The Physical Environment and Its Impact on Human Activities

I need to be able to:	Column A I completed the following:	Column B How did I do? 1. I understand.	2. I somewhat understand.	3. I don't understand.
Develop working definitions for and use correctly the following concepts: • continent • island • ocean • sea • lake • river • bay • gulf • peninsula • isthmus • archipelago.				
Recall the meaning for the terms: grid, latitude, longitude, great circle, hemisphere.				
Use various criteria to classify landmasses and water bodies.				
Use mathematical skills to approximate the proportion of landmasses to water bodies on Earth's surface.				

I need to be able to:	Column A I completed the following:	Column B How did I do?		
		1. I understand.	2. I somewhat understand.	3. I don't understand.
Differentiate between landforms and water bodies (continent, island, ocean, sea, lake, river, bay, gulf, peninsula, isthmus).				
State the absolute and relative location of landforms and water.				
Use lines of latitude and longitude to locate places and features in the world.				
Identify and name lines of latitude and longitude on a map of the world.				
Examine different sources to determine the characteristics of lines of latitude and longitude.				
Create thematic map showing the name and location of the continents, major rivers and oceans of the world.				
Name and describe the parts of the river.				
Work independently to complete individual tasks.				

Objective: Develop working definitions for and use correctly the following concepts: continent, island, ocean, sea, lake, river, bay, gulf, peninsula, isthmus and archipelago.

Item Types: Table grid, single selected and short constructed responses

1. Use the words from the word bank below to complete the table below by adding missing definitions or words.

continent, island, ocean, sea, lake, river, bay, gulf, peninsula, isthmus, archipelago

Words	Definitions
Example: lake	***a body of water that is inland and completely surrounded by land***
	the largest landmass on Earth
island	
ocean	
	the smallest inlet of water protected by land
peninsula	

2. Write a paragraph which includes all the following words:

island, isthmus, peninsula, continent, archipelago

____

____

____

continent, island, ocean, sea, lake,
river, bay, gulf, peninsula, isthmus,
archipelago

3. Read the paragraph below and then fill in the blanks using the words below.

Water bodies cover approximately 75% of the Earth. They come in a variety of sizes. The largest body of water is a/an____________. The second largest body of open water is a/an____________. Water that is fully surrounded by land is known as a/an____________. A stream of water that often forms in mountains is known as a/an ____________. ____________ (s) and____________ (s) are similar because they both form inlets of water protected by surrounded lands.

4. What do Jamaica, Hawaii and Japan all have in common?

A. They are all located in the Atlantic Ocean.
B. They are all peninsulas.
C. They are all surrounded by lakes.
D. They are all part of an archipelago.

Objective: Recall the meaning for the terms: grid, latitude, longitude, great circle, hemisphere.

Item Types: Single selected and short constructed responses

Read the passage below and use it to answer item 5.

To locate places on the Earth, people use the global __________system. This system uses imaginary lines called __________and __________. Lines of latitude run east to west and measure the distance north and south of the Equator. Lines of longitude run north and south and measure distance east and west of the Prime Meridian. Both the Equator and Prime Meridian divide the Earth into __________ (s). The shortest distance of two places on Earth forms the __________ (s).

5. Select the answer which BEST completes the paragraph.

A. grid, longitude, latitude, hemisphere, great circle
B. hemisphere, great circle, grid, longitude, latitude
C. grid, longitude, great circle, latitude, hemisphere
D. latitude, hemisphere, grid, great circle longitude

6. Which statement is TRUE about the lines of latitude?

A. Lines of latitude measure degrees north and south of the Equator.
B. All lines of latitude form great circles.
C. Great circles are only formed by lines of latitude.
D. Only the lines of latitude in the northern hemisphere form a great circle.

Objective: Use various criteria to classify landmasses and water bodies.

Item Types: Table grid, short constructed and single selected responses

7. Fill in the table to rank the continents and bodies of water from largest to smallest.

Name of Continent	Rank (1 = largest, 7 = smallest)
Asia	
Europe	
Antarctica	
North America	
South America	
Africa	
Australasia/Australia	

Body of Water	Rank (1 = largest, 3 = smallest)
bay	
ocean	
sea	

8. Use the information below to populate the Venn diagram.

Asia, Cuba, Europe, Jamaica, Barbados, Africa, North America, Puerto Rico, South America, Australia, Antarctica

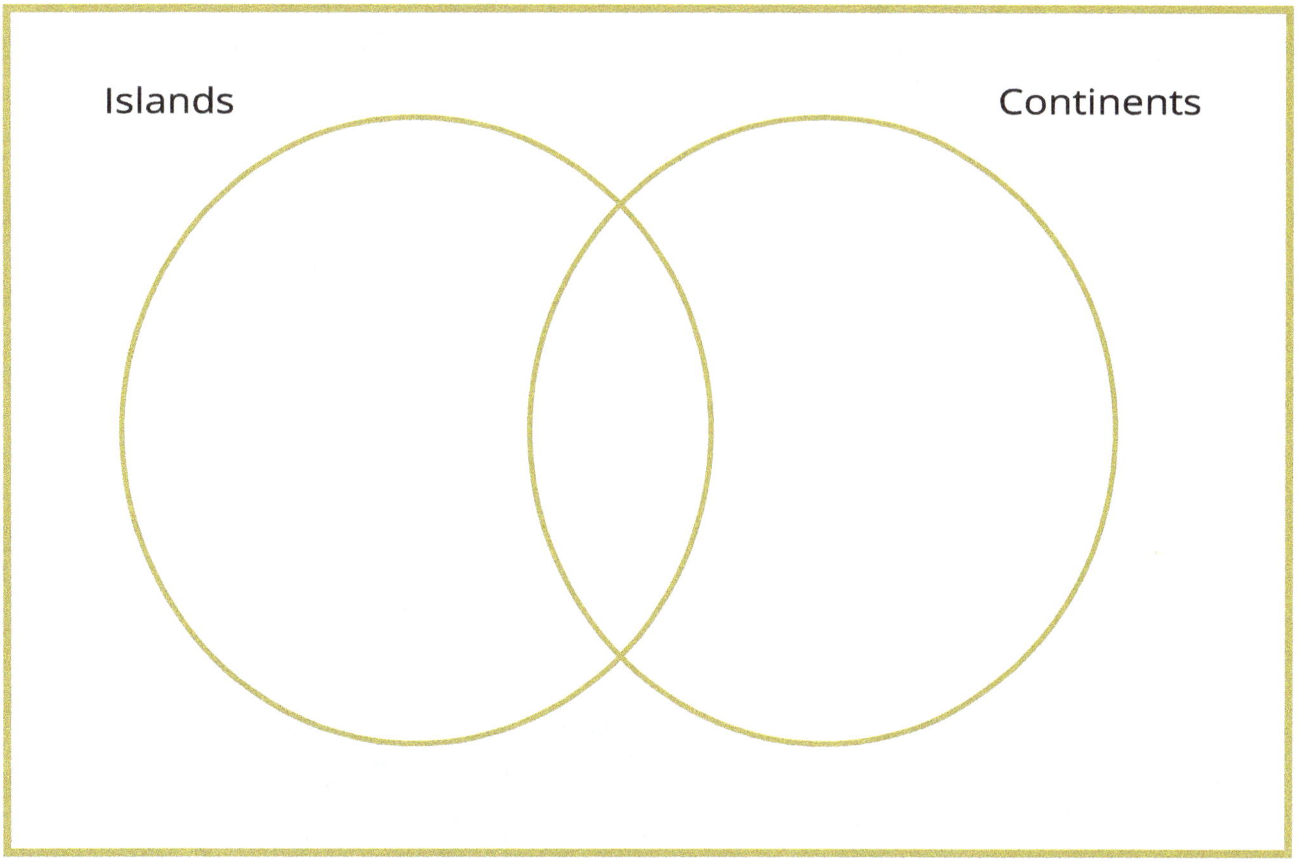

8B. What conclusions can you draw from the Venn diagram?

8C. Water bodies have different characteristics. Use this statement and your knowledge of water bodies to complete the table. Place a tick (✓) under the characteristics that apply in each case.

Water Body	Collects Water	Moves Water	Surrounded by Land	Has no Boundaries	Typically Freshwater
ocean					
sea					
lake					
river					
bay					

9. Find one example of each of the following features on the map below. Label the feature using the letters given.

A. peninsula　C. archipelago　E. ocean　G. lake

B. isthmus　D. island　F. sea　H. continent

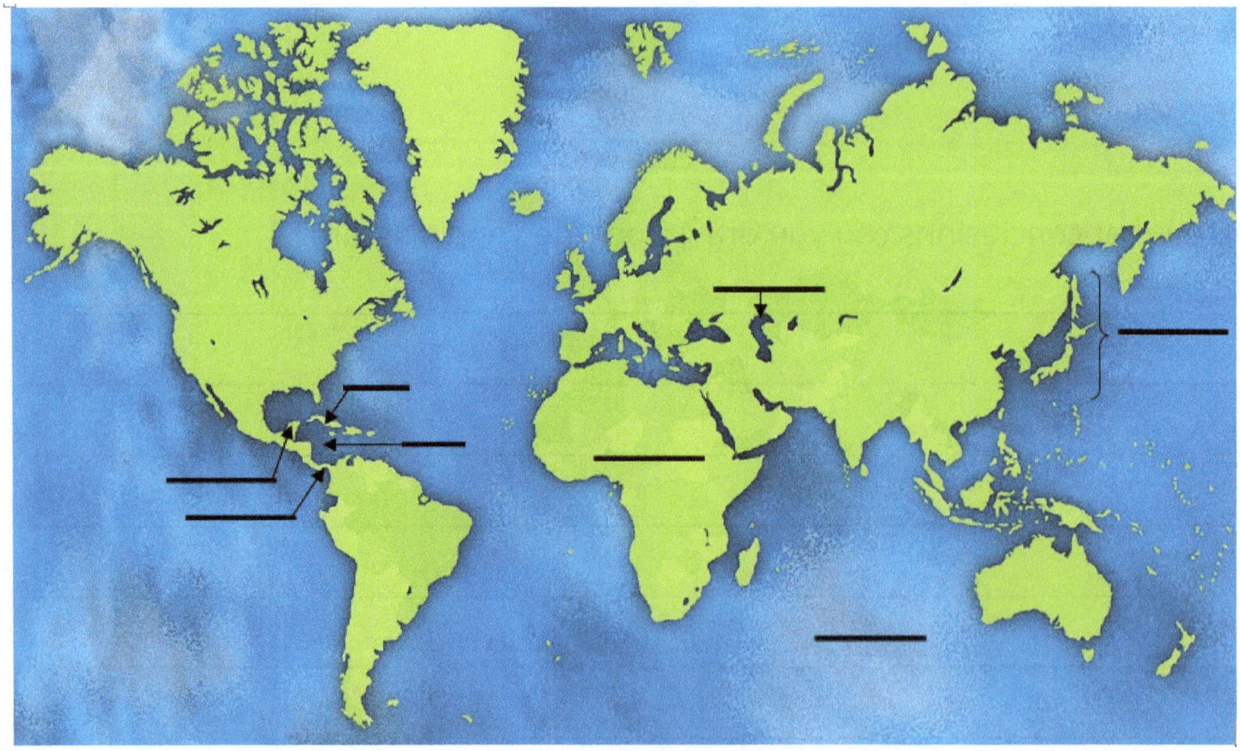

Objective: Use mathematical skills to approximate the proportion of landmasses to water bodies on Earth's surface.

Item Types: Single selected and short constructed responses

Examine the scenario below and answer the question that follows.

> Nicole and Oral were talking about what they learned in social studies class today. Nicole told Oral the landmasses cover more of the Earth's surface than water bodies. Nicole asked Oral to explain why he disagreed. What are some points Oral could make to support his position?

10. Using examples, advise Oral on points to make to explain why he disagreed with Nicole.

____

____

____

____

____

____

____

____

____

____

____

Examine the pie chart and use it to answer item 11.

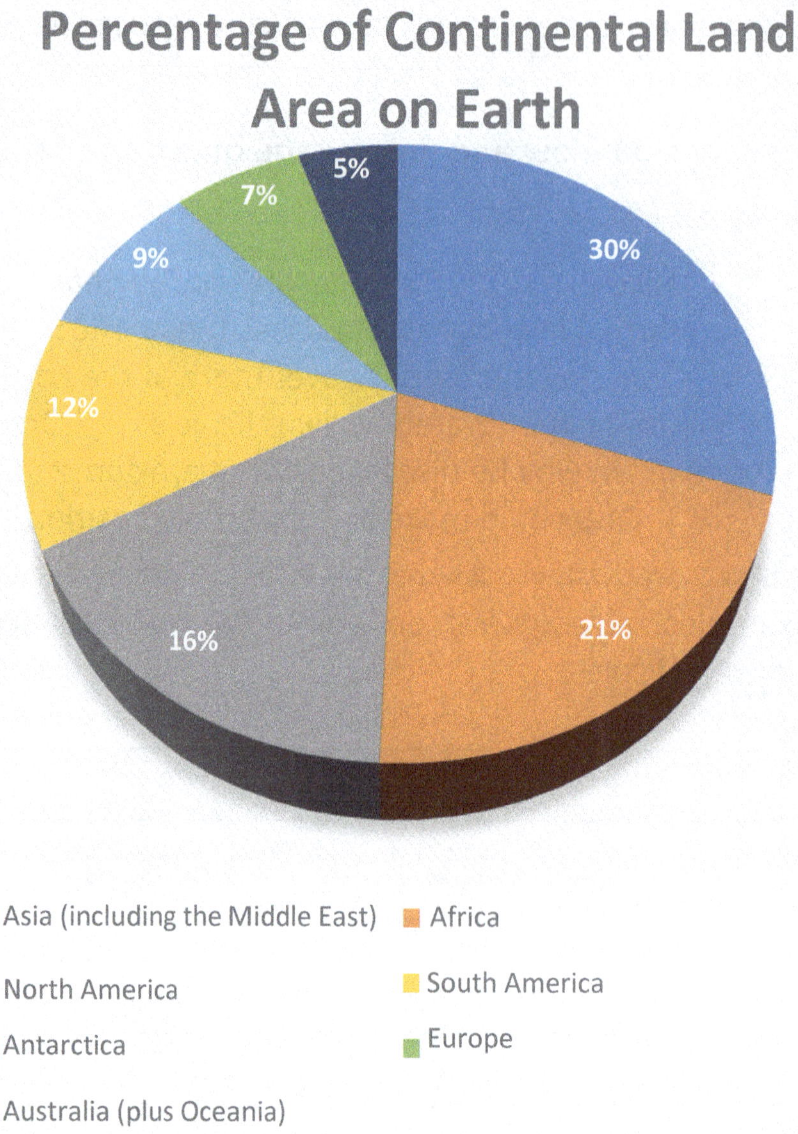

11. Which statement is supported by the pie chart?

A. Asia's and Africa's land areas are more than 50% of the continental land area on Earth.
B. Antarctica's land area is twice as large as Asia.
C. South America's total land area is 22%.
D. Europe's total land area is 16%.

Objective: Differentiate between landforms and water bodies (continent, island, ocean, sea, lake, river, bay, gulf, peninsula, isthmus).

Item Types: Single selected and table grid responses

12. Which statement accurately describes the difference between a bay and sea?

A. A bay is a larger body of water than a sea.
B. A sea is water that is surrounded by land on all sides.
C. A bay and sea have no differences.
D. A sea is a larger body of water than a bay.

Read the table below and use it to answer items 13 and 14.

Sea	River
a large body of water	a long body of running water
made up of salt water	made up of fresh water
formed billions of years ago	can form in a few years
large ships can navigate the waters	most can only handle small ships

13. Which sentence DOES NOT state a difference between seas and rivers?

A. Seas are larger bodies of water than rivers.
B. Rivers can form in only a few years, while seas were formed billions of years ago.
C. Both are bodies of water.
D. Ships of different sizes travel on seas and rivers.

14. Ryan and Crystal are researching landmasses for their test. They realised it was a lot to remember. They created the table below to summarise the most important details about landmasses. Write the information you think they must include in the table.

Landmass	Description of Special Features
isthmus	
peninsula	
continent	
island	

Objective: State the absolute and relative location of landforms and water bodies.

Item Type: Single selected response

Read the passage below. Underline the word from those in brackets which best completes each sentence.

15. Today in school, the teacher told the students that Kingston, Jamaica is located at 18 degrees north and 76 degrees west. The teacher gave them the (absolute, relative) location of Kingston. She then told them that Dunn's River Falls is located on the north side of the island. The teacher gave them the (absolute, relative) location of the falls.

Examine the map and use it to answer item 16.

16. Which statement gives an absolute, then a relative location?

A. Morant Bay is located west of Kingston at 17⁰ north and 75⁰ west.

B. Montego Bay is located at 18⁰ south, 77⁰ west, east of Port Kaiser.

C. Portland Point is north of Portmore, at 17⁰ south and 77⁰ west.

D. Black River is located at 18⁰ north, 77⁰ west.

Objective: Use lines of latitude and longitude to locate places and features in the world.

Item Types: Single selected and table grid responses

Examine the following map and use it to answer items 17 and 18.

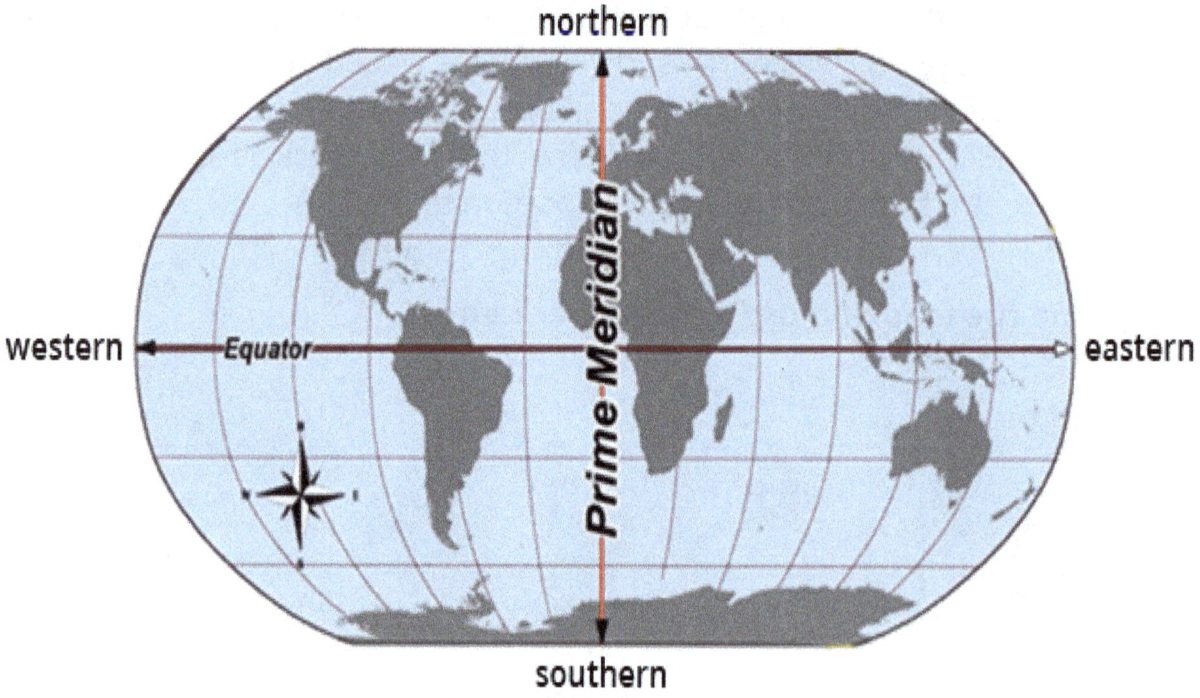

17. To which continent would you need to travel in order to visit the location where the Tropic of Cancer and the Prime Meridian meet?

    A. Asia
    B. North America
    C. Africa
    D. Australia

18. Place a tick (√) beside the accurate statements.

Statement	Tick (√) if accurate
Most landmasses are located south of the Equator.	
The Equator is located between the Tropic of Cancer and the Tropic of Capricorn.	
Jamaica is located between the Equator and the Tropic of Cancer.	
There are no landmasses located south of the Tropic of Capricorn.	

19. The Vice President of the United States of America, is visiting Jamaica. The pilot needs the general coordinates (latitude and longitude) to land in Kingston. Use the map below to give the most accurate coordinate for the location of Kingston.

Source: *www.mapsofworld.com.*
Adapted for educational purposes.

A. 77° 30′ W and 18° 30′ N
B. 18° 30′ N and 76° 30′ W
C. 19° 00′ N and 18° 0′ N
D. 18° 00′ N and 77° 00′ W

Objective: Identify and name lines of latitude and longitude on a map of the world.

Item Type: Single selected response

20. Identify and label the lines of latitude and longitude on the following world map.

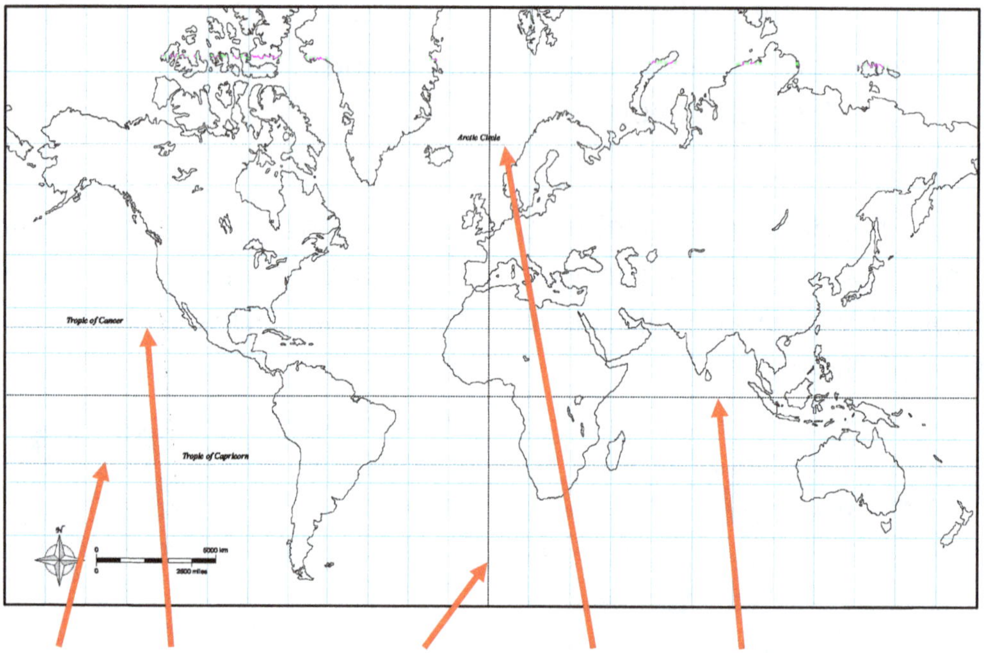

Source: *https://www.worksheetworks.com/geography/world/mercator/coordinate.html.*

Nelson Mandela believed that 'education is the most powerful weapon you can use to change the world.'

Are you ready to learn some more with me so you can change your world?

Come wid mi, mek wi continue!

21A. Read the passage and then select the CORRECT order in which the words should be used to fill in the blank spaces.

> The global grid system is made up of imaginary lines called latitude and longitude. Lines of ________ run east and west on a map, measuring 90 degrees north and 90 degrees south of the ________. Lines of ________ run north and south, measuring 180 degrees east and west of the ________.

A. Equator, longitude, latitude, Prime Meridian
B. Latitude, Equator, longitude, Prime Meridian
C. Prime Meridian, Equator, latitude, longitude
D. Longitude, latitude, Prime Meridian, Equator

21B. Which of the following is not an appropriate use of the lines of longitude?

A. To help determine what time zone a location is in
B. To find a location north of the Equator
C. To locate a place west of the Prime Meridian
D. To find the global grid address of Montego Bay

22. Lines of latitude run north and south of the Equator. These lines cover a total of _______ from the North Pole to the South Pole.

A. 180 degrees
B. 23.5 degrees
C. 90 degrees
D. 360 degrees

Objective: Examine different sources to determine the characteristics of lines of latitude and longitude.

Item Types: Single selected, table grid and short constructed responses

Examine the following sources and use them to answer the questions that follow.

The lands that lie between the Tropic of Cancer and the Tropic of Capricorn are known to have low latitude climates.

"Statement: Locations with a lower latitude number have warmer climates."

23. Do you agree with the above statement? ☐ Yes ☐ No

Provide one example to support your response.

____

____

____

____

____

Use the globes to answer item 24.

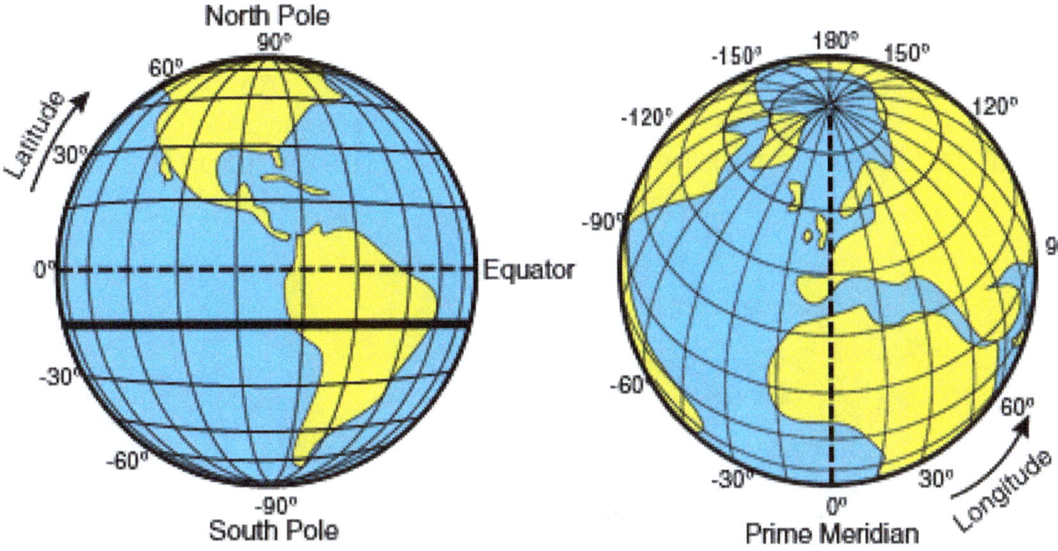

Source: *https://s1.thingpic.com/images/Mj/72gMPG2XvAGLxZoBiGFiWfTp.jpeg.*

24. The chart shows some characteristics of parallel and meridian lines. Complete the chart by writing ONE answer in the blank spaces.

Characteristic	Parallel	Meridians
runs	east to west	
measures		east to west
number		360^0
measure of 0^0	Equator	
distance from each other	equal distance apart	

Objective: Create a thematic map showing the names and locations of the continents, major rivers and oceans of the world.

Item type: Single selected response

25. Locate and label the continents and oceans of the world on the map below.

World map

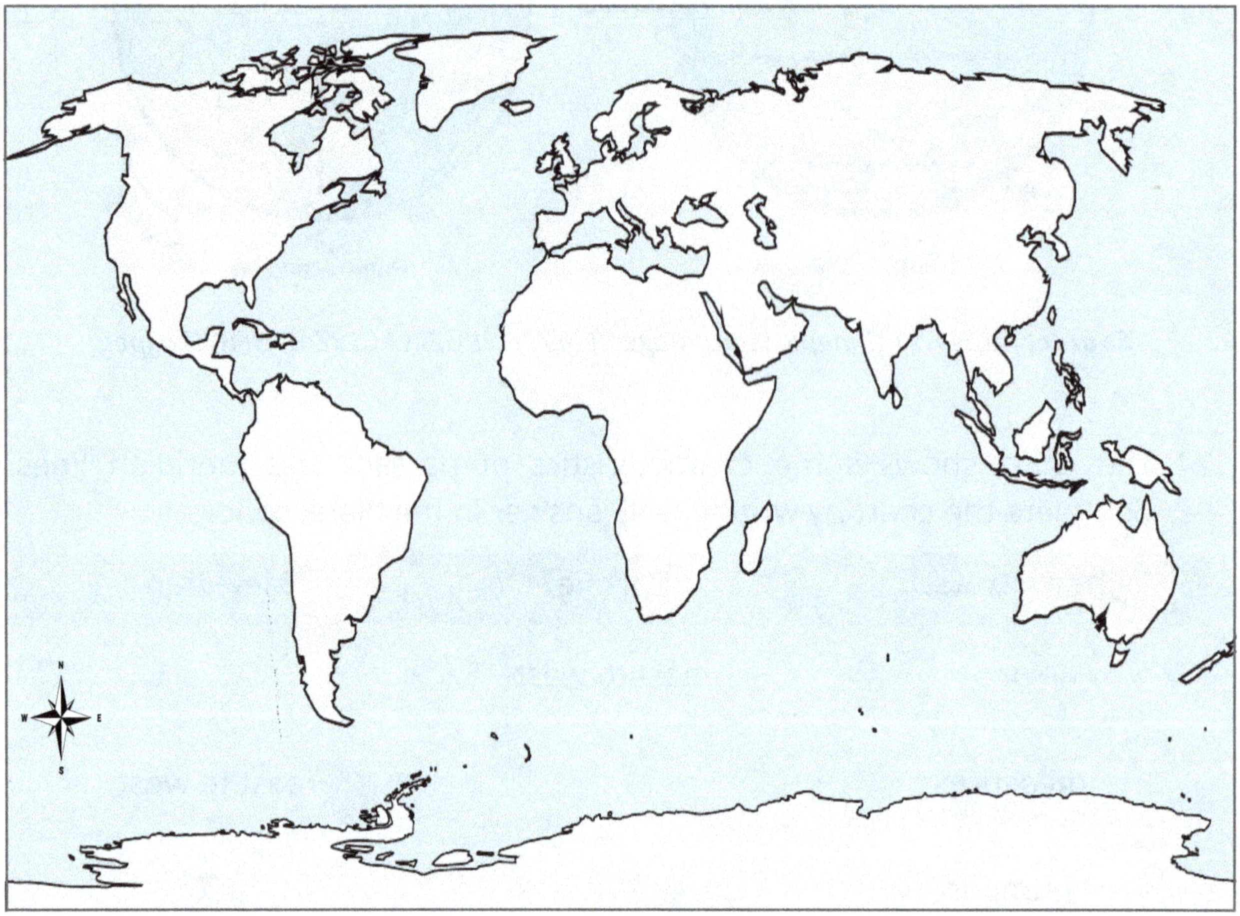

26. Which of these statements is correct?

A. There are five continents and five oceans.
B. There are seven continents and five oceans.
C. There are seven continents and nine oceans.
D. There are nine continents and six oceans.

27. Which river system is NOT correctly paired with the continent on which it is found?

    A. Mississippi River – Europe
    B. Yangtze River – Asia
    C. Nile River – Africa
    D. Amazon River – South America

Source:
https://en.wikipedia.org/wiki/Mississippi_River#/media/File:Efmo_View_from_Fire_Point.jpg.

Objective: Name and describe the parts of the river.

Item Types: Short constructed and single selected responses

28. Label the main parts of the river.

29. Read the paragraph and select the option with the set of words that best completes the paragraph.

River systems are very important parts of the watershed on land. Most rivers start in a mountain range. This is known as the ________. A smaller river that can branch off the main river is called a ________. The sides of the river form the ________. A river reaches the end when it empties into the ________, which is a larger body of water.

A. bank, tributary, mouth, source
B. tributary, mouth, bank, source
C. source, bank, tributary, mouth
D. source, tributary, bank, mouth

Examine the map below and use it to answer item 30.

Map of South America

30. What can the Madeira be classified as?

A. A delta
B. A source
C. A tributary
D. A mouth

Objective: Recognise the usefulness and importance of a geographic coordinate system in solving real-world problems.

Item Type: Single selected responses

31. An earthquake occurs off the coast of Jamaica in the Caribbean Sea. Which statement reflects why it would be important to locate the epicentre using the global grid address?

    A. The global grid address would help find the location where the next island would form.
    B. The global grid address helps the authorities locate the epicenter and issue a tsunami warning to the proper locations.
    C. It would not be possible to find the global grid address because the epicentre was in the Caribbean Sea.
    D. The global grid address would not help because the earthquake occurred hundreds of miles deep in the Earth's crust.

Read the following scenario and use it to answer item 32.

> A cruise ship ran into trouble in the Atlantic Ocean. The ship's captain needed to call the coast guard for help.

32 How would the coast guard be able to find the ship in the vast Atlantic Ocean?

    A. They could find the ship's global address using latitude and longitude coordinates.
    B. The captain could send a smoke signal.
    C. They could use the ocean currents to see where the ship might drift.
    D. They could ask the ship's captain for whatever information they need.

STOP

EVALUATE YOUR WORK!

Remember to evaluate your work by completing the table at the beginning of the unit. You can always review the concepts as well as ask for help.

I understand.	I somewhat understand.	I don't understand.

How was this unit?

I sense that you are really trying.

Be gentle, patient and kind to yourself. Review this section as often as you need. Come back when you are ready. I'll be waiting, but not forever....

Unit Assessment Based on the NSC

Term 2, Unit 3

Focus question: 'How are decisions made at the national level and how do these decisions affect us?'

Source: *https://jis.gov.jm/government/the-legislature/*

'Educate and inform the whole mass of the people... They are the only sure reliance for the preservation of our liberty.'

– Thomas Jefferson

TERM 2, UNIT 3

Living Together

I need to be able to:	Column A I completed the following:	Column B How did I do?		
		1. I understand.	2. I somewhat understand.	3. I don't understand.
Develop working definitions and use correctly the following terms: • citizen • leader • democracy • cabinet • government • parliament • opposition • senate • monarch • constitution • vote • constituency.				
Distinguish between local and central government.				
State the requirements for Jamaican citizenship.				
Participate in activities that foster and develop responsible actions by citizens.				
Distinguish between rights and responsibilities of citizens.				

I need to be able to:	Column A I completed the following:	Column B How did I do? 1. I understand.	2. I somewhat understand.	3. I don't understand.
Examine the rights of a Jamaican citizen and develop a list of responsibilities of a citizen that complement these rights.				
Examine an organisational chart of the structure of the Jamaican system of government, then describe it and make deductions about the relationship among members.				
Identify persons in positions of power, describe how they acquired the position of power and how their use of this power affects the freedoms and rights of others.				
Examine the activities of various leaders, then develop and justify a list of skills and qualities needed to lead at the national level.				
Identify the goods and services provided by government and explain how the government gets money to pay for these.				
Evaluate various decisions made by the Jamaican				

I need to be able to:	Column A I completed the following:	Column B How did I do? 1. I understand.	 2. I somewhat understand.	 3. I don't understand.
government and discuss the intended and unintended impacts of these decisions on the Jamaican people, then propose amendments to the decisions.				
Compare the procedures for making decisions in various settings (classroom, school, home, community, government).				
Examine various cases of how justice is meted out to citizens of Jamaica, then develop criteria to judge the degree of fairness and use them to evaluate cases and propose just measures.				
Work cooperatively and individually to accomplish goals.				

Source: *http//:www.jis.gov.jm/portiaXedit-670x450.*

Figure 30: *Past and current Prime Ministers hugging each other.*

Objective: Develop working definitions and use correctly the following terms: citizen, leader, democracy, cabinet, government, parliament, opposition, senate, monarch, constitution, vote, constituency.

Item Types: Single selected and order match responses

Read the following passage and use it to answer item 1.

Every country has a form of government, a group of people who run the country and make rules and laws for the citizens. Citizens are the people who live in the country. Jamaica's government is a constitutional democracy. In Jamaica, eligible citizens vote for their council leaders. To vote means to cast a ballot for people who will make the laws. The leaders are chosen to represent people who are from their constituency.

There are several types of governments. In a democracy, the people have the right to vote for their leaders and laws. Constitutional democracies have a constitution, that is, a guideline for the structure of the government and rules that the government follows. A country ruled by a king or queen is known as a monarchy. Its ruler is called a monarch.

Governments have different types of governing bodies. A parliament is a law-making group that represents the people and oversees the workings of the government. A senate is another type of legislative body that sometimes has more power than a lower body of representatives. A cabinet is a small group of officials who assist the leader of the council more directly. A group of people who form part of the government but are not currently in power are called the opposition.

Source: *https://jis.gov.jm/ministers-and-attorney-general-take-oath-of-office.*

Figure 31: *Fourteen Ministers of Government and the Attorney-General were sworn into office on Sunday (September 13, 2020), during a ceremony at King's House.*

1. The table below contains a list of words in column A and matching definitions in column B. In Column A, write the letter for the corresponding definitions on the lines given. Each word should be used only once.

COLUMN A	COLUMN B
___ **citizen**	**A.** a division of the country where the people are represented by a certain person in the government
___ **leader**	**B.** a legal national of a country
___ **democracy**	**C.** a group of people who ensure that the party in power does not act against the best interests of the people
___ **cabinet**	**D.** to cast a ballot for a person running for office
___ **government**	**E.** someone responsible for guiding a group, institution, country, etc.
___ **parliament**	**F.** a written set of rules on the organisation of the government and the rights and duties of citizens
___ **opposition**	**G.** a country ruled by a king or queen
___ **senate**	**H.** a group of lawmakers who represent the people and oversee the government
___ **monarchy**	**I.** a small group of special representatives who closely assist the leader
___ **constitution**	**J.** a group of people who run the country and make rules and laws that govern the country
___ **vote**	**K.** a government that allows the citizens to vote for their representatives
___ **constituency**	**L.** a legislative body that has a little more power than a lower body

Read the passage carefully, then select the correct order in which the following words should appear to complete the paragraph accurately. Each word can only be used once.

democracy	constitution	monarch	government

Every country has persons who lead and make decisions. They form what is known as a ____________. Many governments have a certain set of rules and structure written down in a_____________. Citizens have the right to vote for their representatives in a ________. A ruler who is a king or queen is known as a __________.

2. Which of the following is the correct choice for the passage above?

A. Government, monarch, constitution, democracy
B. Democracy, government, constitution, monarch
C. Government, constitution, democracy, monarch
D. Constitution, democracy, monarch, government

Objective: Distinguish between local and central government.

Item Types: Short constructed, table grid and single selected responses

Read the following scenario and use it to answer item 3.

Richie is upset that the hurricane season is approaching and the drain in his community has not been cleaned. His house was flooded during the last rainy season and he fears that this will happen again. He wants to write to the Prime Minister for help to get the drain cleaned.

3. Explain to Richie what the main difference is between local and central government and who he should appeal to for help.

____

____

____

____

____

____

____

____

____

____

____

____

____

____

____

4. Look at the roles listed below and use a tick (✓) to indicate if they are carried out by central or local government.

Roles	Central Government	Local Government
They create policy on matters such as health, education, finance.		
They provide help for the poor, homeless, elderly in communities.		
They implement and change laws in the country's best interest.		
They maintain street lights.		
They are responsible for beaches and public parks.		

5. Which statement is TRUE about the main difference between how local and central governments are elected?
   A. The people of a parish have no say in electing their officials, while the leaders of the central government elect representatives to the political party.
   B. The people of a parish directly elect their council representatives, while the head of the central government is chosen by a political party.
   C. The people of Jamaica elect all leaders in both the local and central government.
   D. A single political party runs the local government, while several political parties run the central government.

Objective: State the requirements for Jamaican citizenship.

Item Types: Single selected and multiple selected responses

6. Which of the following is NOT a way to obtain Jamaican citizenship?

   A. By birth
   B. Through marriage to a citizen
   C. By living in Jamaica and contributing to its development for at least five years
   D. By being in possession of a valid work permit

7. Which TWO statements are TRUE about the requirements for becoming a Jamaican citizen?

   A. Any baby who is adopted illegally can become a citizen.
   B. Any person born in Jamaica after August 5, 1962, is automatically a Jamaican citizen
   C. Any person born in the British commonwealth is a Jamaican citizen.
   D. Anyone born outside of Jamaica to parents who are citizens of Jamaica at the time of birth is a Jamaican citizen.

Objective: Participate in activities that foster and develop responsible actions by citizens.

Item Types: Single selected and table grid responses

8. Which statement shows how citizens can participate in activities that foster and develop responsible actions?

   A. Do well in school to make your parents, teachers and community proud.
   B. Support police officers who make requests for lunch or drink money as a way of showing respect for their hard work.
   C. Start a community action group to address the need for more recycling in an area.
   D. Attend all council meetings and report persons who do not attend.

9. Selena needs a better understanding of specific activities she can do to show responsible citizenship.

Tick (√) the activities that show how to become a responsible citizen.

Activity	Supports Becoming a Responsible Citizen
Serve on a jury.	
Have multiple unpaid traffic tickets	
Volunteer to help disabled children.	
Form a group that would promote a law to compel everyone in their communities to join the local community club.	

Objective: Distinguish between rights and responsibilities of citizens.

Item Type: Single selected response

Read the following sources and use them to answer item 10.

Source 1	Source 2
________________	________________
• They cannot be arrested without a cause. • They can assemble peacefully. • They are free to join any church they want.	• They should pay taxes. • They should obey the laws. • If they witness a crime, they should serve as witness in court if asked.

10. Which statement shows the correct heading for each source above?

A. Source 1 – Rights; Source 2 – Responsibilities
B. Source 1 – Responsibilities; Source 2 – Rights
C. Source 1 – Prohibited; Source 2 – Allowed
D. Source 1– Legislative powers; Source 2 – Executive powers

11. Read the statement below and then select the most appropriate response to inform Jonathan.

> Jonathan has the right to vote in the upcoming parish council election, but he decides to go to the beach to help clean the shoreline instead. His friend, Parker, suggests that he is committing a crime and will get into trouble by not voting.

A. Jonathan cannot be arrested because cleaning the beach is an important responsibility for citizens to carry out.

B. Jonathan is breaking the law and will spend six months in jail If he does not vote

C. Jonathan has the right to vote, but it is his voluntary responsibility to vote, so he is not breaking the law.

D. Jonathan is responsible as a citizen because it is more important to take care of the environment than voting for your leaders.

Source: *http//:www.jis.gov.jm/o-BALLOT-BOX-facebook-670x450.*

Objective: Examine the rights of a Jamaican citizen and develop a list of responsibilities of a citizen that complement these rights.

Item Type: Table grid response

12. Examine the rights of Jamaican citizens in the column on the left, then write a responsibility that the citizens have that complements each right.

Rights	Responsibilities
right to life	
not being arrested without a cause	
right to have your own opinion on matters	
the right to assemble peacefully	
protection from discrimination	
the right to free speech	
the right to a fair trial	
the right to vote	
the right to worship in any religion	
the right to freedom of movement	

Objective: Examine an organisational chart of the structure of the Jamaican system of government, then describe it and make deductions about the relationship among members.

Item Type: Single selected responses

Examine the organisational chart below and use it to answer items 13 and 14.

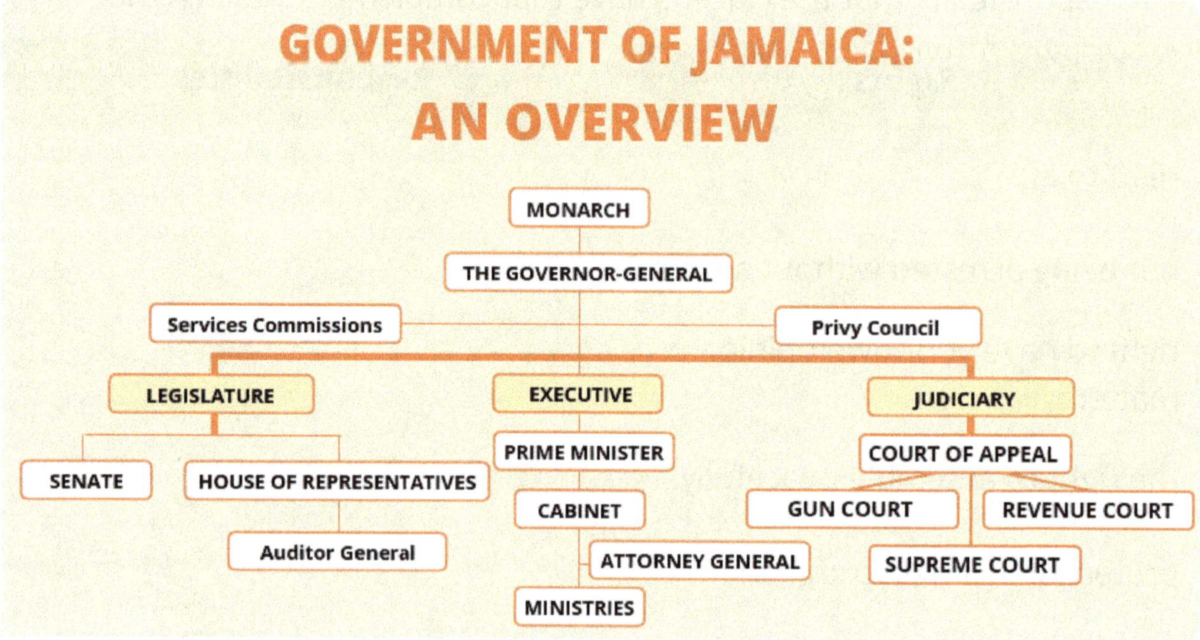

Source: *https://jis.gov.jm/features/overview-government-jamaica.*

13. Which statement is TRUE about the relationship between the Prime Minister and the Attorney General?
    A. The Prime Minister has more authority than the Attorney General.
    B. The Prime Minister has less authority than the Attorney General.
    C. The Prime Minister has the same authority as the Attorney General.
    D. The Prime Minister has no authority while the Atorney General has all the authority.

14. What are the THREE arms of government in the diagram?
    A. senate, house of representative, cabinet
    B. legislature, executive, judiciary
    C. Prime Minister, cabinet, ministries
    D. monarch, the Governor General, Prime Minister

Objective: Identify persons in positions of power, describe how they acquired the position of power and how their use of power affects the freedoms and rights of others.

Item Types: Table grid and short constructed responses

15A. Identify THREE positions or job titles within our current government. Outline the process by which they acquired the position. Use this information to complete the table.

Name	How Position is Acquired
Example: The Governor General	**appointed by the monarch, whom he represents as head of state in ceremonies**
1.	
2.	
3.	

15B. For TWO of the positions or job titles you have selected, say how their use of power affects the rights and freedom of others.

____

____

____

____

____

Objective: Examine the activities of various leaders and then develop and justify a list of skills and qualities needed to lead at the national level.

Item Type: Short constructed responses

16. Identify TWO skills and TWO qualities a Prime Minister should have to be an effective leader of a country.

Skills and Qualities

17. Explain why you believe these skills and qualities are needed by a leader of a country.

____

____

____

____

____

____

____

____

Objective: Identify the goods and services provided by government and explain how the government gets money to pay for these.

Item Types: Table grid and multiple selected responses

18A. Below is a list of the current ministries found in Jamaica. Using what you know about the goods or services that each ministry provides, complete the table below.

Ministry	One Example of Goods or Services
Culture, Gender, Entertainment and Sport	
Economic Growth and Job Creation	
Health And Wellness	
Justice	
National Security	
Science, Energy and Technology	
Transport and Mining	

18B. Select THREE ways that the government collects revenue to provide these goods and services. Taxes ar collected from:

A. Goods and services sold
B. Employees' salaries
C. Health insurance policies
D. Property owners

Objective: Evaluate various decisions made by the Jamaican government and discuss the intended and unintended impacts of these decisions on the Jamaican people, then propose amendments to the decisions.

Item Type: Short constructed responses

Read the paragraph below and use it to answer items 19 to 21.

In March 2020, the government of Jamaica decided to close schools to reduce the number of COVID-19 infections and deaths that were occurring. Some schools were able to transition to online teaching and learning. Other schools used distance education methods, such as sending books and materials to children for them to complete and submit.

19. Give your opinion on the government's decision to close schools.

____

____

____

20. Explain ONE intended and ONE unintended impact of the decision.

____

____

____

21. If you could advise the government on a change that would improve the situation, what would you suggest?

____

____

____

Objective: Compare the procedures for making decisions in various settings (classroom, school, home, community, government).

Item Type: Extended writing constructed response

Complete this exercise in small groups. Read the passage below and answer item 22.

Patrice was very excited when she passed her PEP exams for the school of her choice. She is now in Grade 8, is on the honour roll, participates in several clubs and has made some wonderful friends. Being at the school is a dream come true for her. Her brother, Judah, is also very active at school and is the captain of the quiz team.

Yesterday, their parents called a family meeting. They found a lovely home that they want to buy in the country. This will mean that Patrice and Judah will have to transfer to new schools. Judah doesn't mind. He has been sleeping in the living room because the small apartment only has two bedrooms. Moving is not something that Patrice is interested in. Her parents believe this is a good move for the family. Housing in the city is very expensive and they do not want to continue paying higher rent when they could own a home elsewhere. They are considering Patrice's objection. They will let the children know their final decision by the end of the week.

22A. List the steps taken in the family's decision-making process.

22B. Choose another organisation, such as a school or service club and create a diagram showing what you think their decision-making process is.

22C. Work in a group to discuss the similarities and differences with the decision-making processes presented in 22A and 22B. Write your conclusions in the space provided.

Objective: Examine various cases of how justice is meted out to citizens of Jamaica, then develop criteria to judge the degree of fairness and use these to evaluate cases and propose just measures.

Item Type: Single selected response

Read the following statement and use it to answer items 23.

> Some citizens claim to have been abused by members of the Jamaica Constabulary Force (JCF). Independent Commission of Investigation (INDECOM) is investigating the matter.

23. Which of the following would be a good measure to take to protect the citizens against such abuse?

   A. The government should provide an organization to counsel people who have been abused by the police.
   B. Institute a law mandating all members of the constabulary force to go to church.
   C. Set up an independent body to investigate and hold JCF officers accountable if they are accused of violating citizens' rights.
   D. Provide rewards for police who abuse citizens.

24. Mr. Smith was arrested and charged for not paying his taxes. In which of the following courts would his case be tried?

   A. Traffic court
   B. Gun court
   C. Revenue court
   D. Family court

According to Jack Lew, 'I think there's no higher calling in terms of a career than public service, which is a chance to make a difference in people's lives and improve the world.'
What do you think? Come wid mi! Mek wi continue.

SOCIAL STUDIES TEST

CHEETAH 40-QUESTIONS TEST

ANSWER SHEET #1

Use the answer sheet to record your answers. Shade in the circle that corresponds to your answer for each test item.

Name: ______________________ **ID no:** ____________ **Age:** ____

School: ____________________ **School address:** ____________________

1	A	B	C	D	21	A	B	C	D
2	A	B	C	D	22	A	B	C	D
3	A	B	C	D	23	A	B	C	D
4	A	B	C	D	24	A	B	C	D
5	A	B	C	D	25	A	B	C	D
6	A	B	C	D	26	A	B	C	D
7	A	B	C	D	27	A	B	C	D
8	A	B	C	D	28	A	B	C	D
9	A	B	C	D	29	A	B	C	D
10	A	B	C	D	30	A	B	C	D
11	A	B	C	D	31	A	B	C	D
12	A	B	C	D	32	A	B	C	D
13	A	B	C	D	33	A	B	C	D
14	A	B	C	D	34	A	B	C	D
15	A	B	C	D	35	A	B	C	D
16	A	B	C	D	36	A	B	C	D
17	A	B	C	D	37	A	B	C	D
18	A	B	C	D	38	A	B	C	D
19	A	B	C	D	39	A	B	C	D
20	A	B	C	D	40	A	B	C	D

Score ____________ **out of 40**

This page can be used to write notes.

TEST #1

Answer the questions below by selecting the correct responses.

1. Which group was the first set of immigrants to arrive in the Caribbean region?

   A. Africans
   B. Amerindians
   C. Asians
   D. Europeans

2. Which TWO of the following definitions best describe the term indentureship?

   A. a contract of employment for a set period
   B. wages for work done for a period of time
   C. a legal agreement between two parties outlining conditions for a job
   D. a passage from India to the Caribbean to work

Read the passage and then answer question 3.

This aspect of culture was brought to the Caribbean by an ethnic group in the 19th century. Families come together to celebrate in different ways. They listen to stories about different gods and wear colourful costumes. Lamps are lit in the homes and on the streets. Special foods are eaten and a parade is held.

3. To which ethnic group and celebration does the passage refer?

   A. Chinese, New Year
   B. Portuguese, Eid al-Fitr
   C. Africans, Kumina
   D. Indians, Diwali

4. Which is NOT true of the conditions experienced by the East Indian labourers when they were brought to the Caribbean?

   A. They were given housing.
   B. They were given basic food rations.
   C. They were given clothing and tools.
   D. They were free to leave the plantations at any time.

5. Which TWO Caribbean countries gained independence in the same year?

   A. Cuba and Haiti
   B. Jamaica and Barbados
   C. Haiti and Trinidad
   D. Jamaica and Trinidad and Tobago

6. When a country gains independence, which of these will NOT be affected?

   A. National symbols
   B. Constitution
   C. Location
   D. Currency

7. St. Helen's Primary School has had a national flag for over twenty-five years. It is now faded, so the board decided to discard it and replace it with a new one. Which of the following protocols should be observed when disposing of the Jamaican flag?

   A. Throw it in the bin.
   B. Bury it in a public ceremony.
   C. Burn it privately.
   D. Cut it up and recycle it.

8. The national anthem is a prayer. In the line 'Guard us with Thy mighty Hand' what is being prayed for?

   A. Provision
   B. Protection
   C. Assistance
   D. Health

Examine the poster below and then answer question 9.

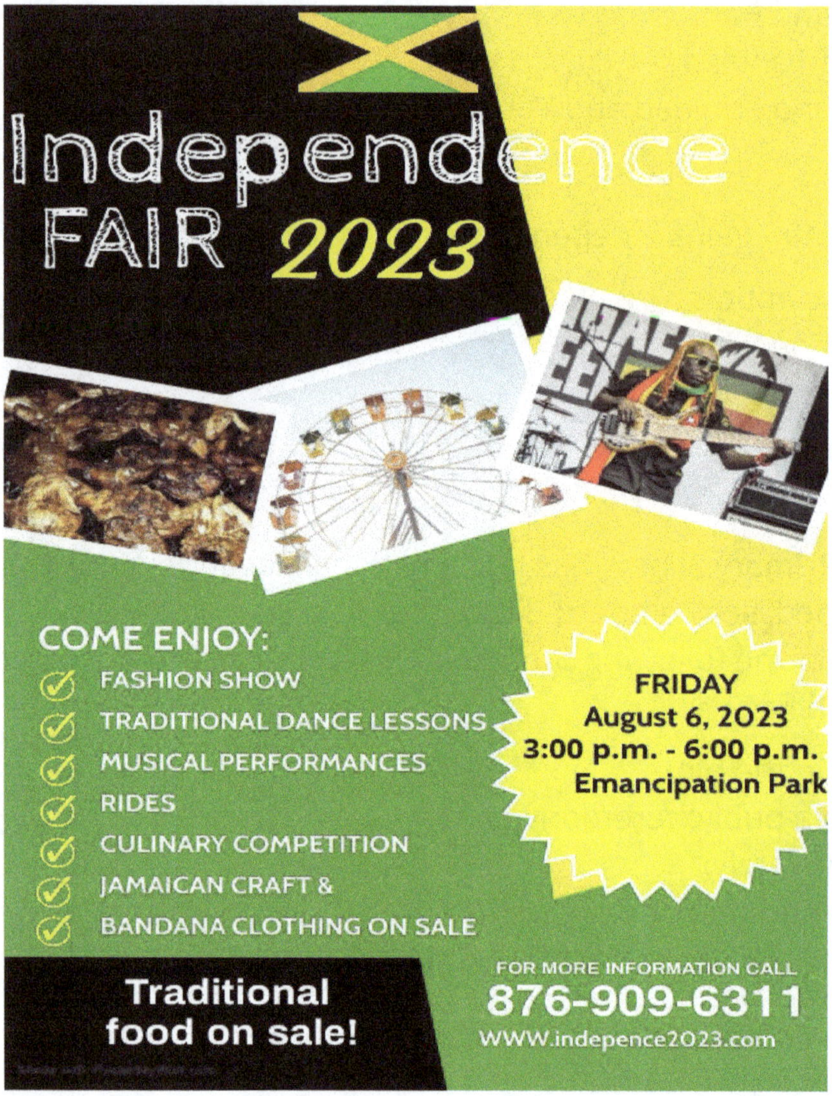

9. From the description of the poster above, choose the option that shows only goods.

   A. Jamaican craft, traditional food, bandana clothing

   B. culinary competition, rides, traditional dance lessons

   C. fashion show, traditional food, traditional dance lessons

   D. musical performances, bandana clothing, culinary competition

10. Choose the option with the correct dates of each event arranged in chronological order.

A. Universal adult suffrage (1944), Morant Bay Rebellion (1865), Independence (1838), Emancipation (1962)

B. Independence (1962), Morant Bay Rebellion (1944), Emancipation (1838), Universal adult suffrage (1865),

C. Emancipation (1838), Morant Bay Rebellion (1865), Universal adult suffrage (1944), Independence (1962)

D. Morant Bay Rebellion (1838), Emancipation (1865), Universal adult suffrage (1944), Independence (1962)

11. Choose the national hero who founded the Universal Negro Improvement Association (UNIA).

A. Norman Washington Manley

B. Alexander Bustamante

C. Marcus Mosiah Garvey

D. George William Gordon

12. Choose the best description for the Jamaican coat of arms.

A. It shows two Taino Indians, with a green cross and the Jamaican iguana.

B. It shows a crocodile with a Taino family and a basket of Jamaican *Lignum vitae*.

C. It shows a helmet, the Jamaican motto and five mangoes forming an X.

D. It shows two Taino Indians, a shield with five pineapples and a crocodile.

13. Choose the option that has words which could be inserted to correctly complete the paragraph.

Jamaica is made up of various ________________ that ________________ to the country during the period of ________________ as well as after ____________________. While the majority of the population is of ______________________descent, the diverse mix of nationalities creates a melting pot of culture, giving true meaning to the motto, 'Out of Many, One People.'

A. emancipation, African, immigrated, colonisation, ethnic groups
B. ethnic groups, immigrated, emancipation, colonisation, African
C. ethnic groups, immigrated, colonisation, emancipation, African
D. African, immigrated, emancipation, colonisation, ethnic groups

14. Examine the illustration below and select the group that correctly identifies the landforms labelled X, Y and Z.

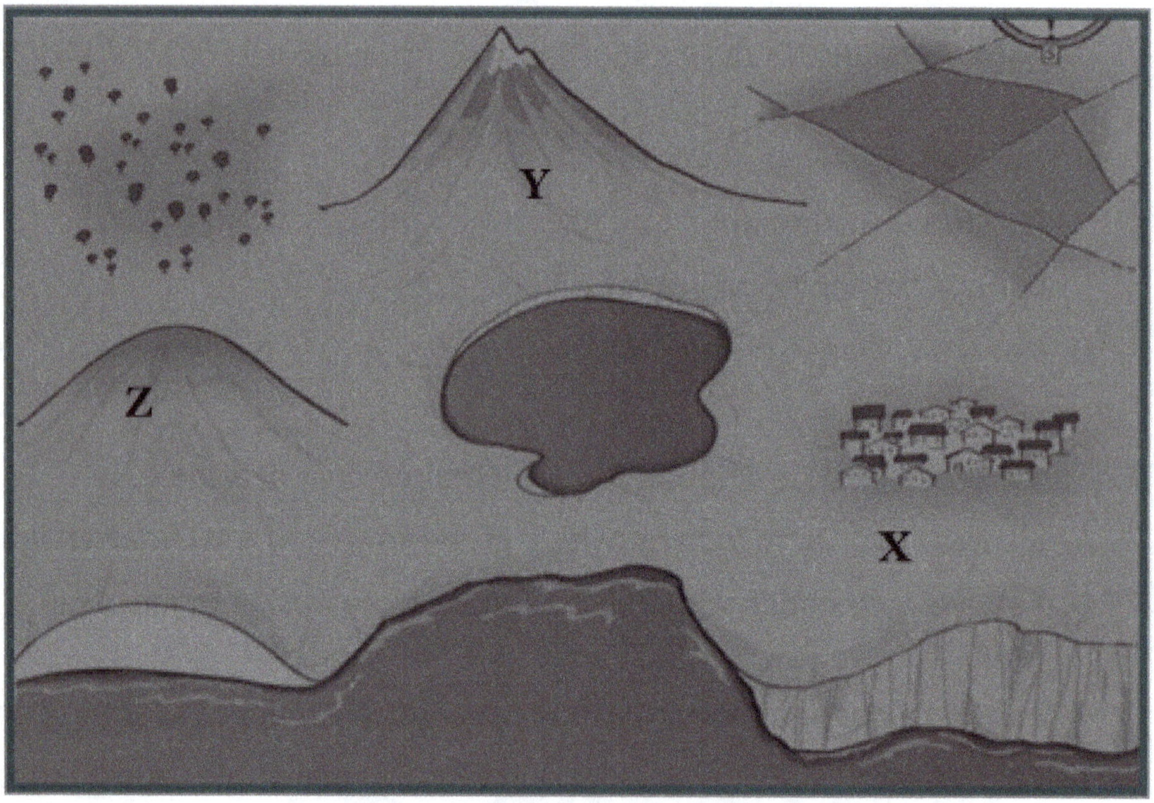

A. X – mountain, Y – hill, Z – plain
B. X – plain, Y – mountain, Z - hill
C. X – hill, Y – valley, Z – mountain
D. X – valley, Y – hill, Z – plain

Read the statement below and then respond to questions 15, 16 and 17.

'The Bellevue Peak is 1151 metres (1151 m) above sea level, Catherines Peak is 1353 m above sea level, Blue Mountain Peak is 2257 m above sea level and the Candle Fly Peak is 1505 m above sea level.'

15. On which of these mountains will the temperature be the lowest?

    A. Candle Fly Peak

    B. Catherines Peak

    C. Blue Mountain Peak

    D. Bellevue Peak

16. Explain why the lowest temperature will be experienced on this mountain.

    A. It is a parish that is very cold.

    B. It gets the most rainfall.

    C. It has the highest elevation.

    D. It has the lowest elevation.

17. Rank the mountains in ascending order (lowest to highest)

    A. Candle Fly Peak, Bellevue Peak, Blue Mountain Peak, Catherines Peak

    B. Bellevue Peak, Catherines Peak, Candle Fly Peak, Blue Mountain Peak

    C. Blue Mountain Peak, Candle Fly Peak, Catherines Peak, Bellevue Peak

    D. Bellevue Peak, Candle Fly Peak, Catherines Peak, Blue Mountain Peak

18. Which of the following statements are true?

I. Lines of longitude run from north to south.

II. All the lines of latitude form a great circle.

III. The Tropic of Cancer is a line of latitude.

IV. The Prime Meridian cuts the Earth into the Northern and Southern Hemispheres.

A. I, III,

B. I, II, III

C. IV, II, I, III

D. IV, I,

Use the diagram below to answer questions 19, 20 and 21.

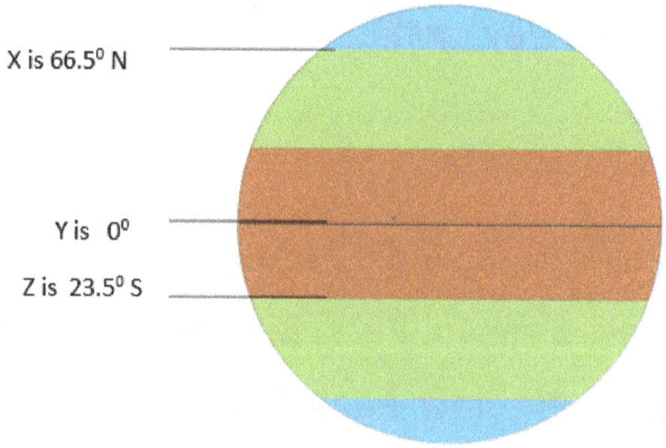

19. Name the line labelled X.

A. Tropic of Capricorn

B. Arctic Circle

C. Equator

D. Tropic of Cancer

20. Name the line labelled Y.

A. Tropic of Capricorn
B. Prime Meridian
C. Equator
D. Tropic of Cancer

21. Name the line labelled Z.

A. Tropic of Capricorn
B. Arctic Circle
C. Equator
D. Antarctica

Read the passage and then answer questions 22 and 23.

Rushell is researching the arrival of indentured labourers in the Caribbean between 1854 and 1900. She wants to know about the total population on each trip, the difference in the number of males and females and the ratio of Indians to Chinese.

22. Which source is the most suitable to provide Rushell with the information needed?

A. Interview with Chinese immigrants
B. Ship records between 1845 and 1916
C. Newspaper articles from 1870
D. The National Library of Jamaica

23. While examining documents during her research, Rushell finds a copy of an emigration pass issued to Raj Maragh in 1857. What information would most likely be found on this pass?

i. Name
ii. Schools attended
iii. address
iv. Favourite food
v. Destination
vi. gender
vii. Country of origin
viii. Foot size

A. i, ii, iii and iv
B. i, iii, v, vi and vii
C. i and viii
D. all of the above

24. Removal of trees can negatively impact mountainous environments. Which TWO of the following are possible environmental effects of this activity?

A. Soil erosion
B. Decrease in land value
C. Unattractiveness of the landscape
D. Loss of habitats for animals

25. Which TWO responses below would support the statement, 'The indentured labourers have made significant contributions to the development of Jamaica.'

A. The mixture of skin colour
B. Their cultural practices
C. Business prowess
D. Hosting of festivals

26. The statement below was made by a black farmer in St. Thomas in 1944.

'Finally! We are now able to vote. I will be first in the line on election day. No longer can I be prevented from having my voice heard about the running of my country.'

To which historical event does his statement refer?

A. Independence
B. Emancipation
C. Universal adult suffrage
D. Christmas Rebellion

27. You were not born in Jamaica, but you have lived in the country for the past five years. However, you still have not yet applied for citizenship. Select TWO ways in which you will be affected as a resident in the country?

Inability to ____.

A. Own a Jamaican passport
B. Go to school
C. Vote in a general election
D. Buy Jamaican products

28. The central government is divided into three arms which work together to ensure that the country is operating at its best. These three arms are the legislature, the executive and the judiciary. The Member of Parliament for North West Clarendon is heading to his weekly meeting to discuss the construction of bridges across the island. With which of these groups is he most likely to meet?
    A. Senate
    B. House of representatives
    C. Parish council
    D. Courthouse

29. There were conflicts on the sugar plantations between the Chinese and the formerly enslaved Africans. However, after the Chinese left the sugar plantations, the relationship improved.
    Identify TWO factors that may have led to this conflict.
    A. The Chinese were migrants.
    B. The formerly enslaved Africanssaw them as threats.
    C. The Chinese were paid less than the formerly enslaved Africans.
    D. The formerly enslaved Africans bossed the Chinese around.

30. Johanah called her friend in England. She realised that her friend was sleeping because it was 3:20 a.m. in England and it was 8 p.m. where Johanah was.
    Which of the following pairs of concepts can be used to explain this?
    A. Condensation and evaporation
    B. Precipitation and hallucination
    C. Rotation and revolution
    D. Orbit and colonisation

Read the facts about Nadia in the box below. Use them to respond to items 31 and 32.

- Nadia is twelve years old.
- She lives in a colony in the Caribbean.
- The official language of the colonizers of the country

31. Which of these countries is likely to be Nadia's homeland?

A. Anguilla
B. Aruba
C. Puerto Rico
D. The Cayman Islands

32. Which of the following statuses would the colony mentioned above have?

A. Independent
B. Colonizer
C. Dependent
D. Overseas region

33. Complete this sentence with an option from the list below:

A country that is colonised

A. Manages the affairs of another country.
B. Is settled in and developed by another.
C. Is taken and governed by another country.
D. Is sold by another country.

34. Which pair of countries would have a high commissioner in Great Britain?

A. Jamaica and Barbados
B. Antigua and Venezuela
C. Hispaniola and Haiti
D. Trinidad and Tobago and Cuba

Examine the organisational chart below and then answer questions 35 to 37.

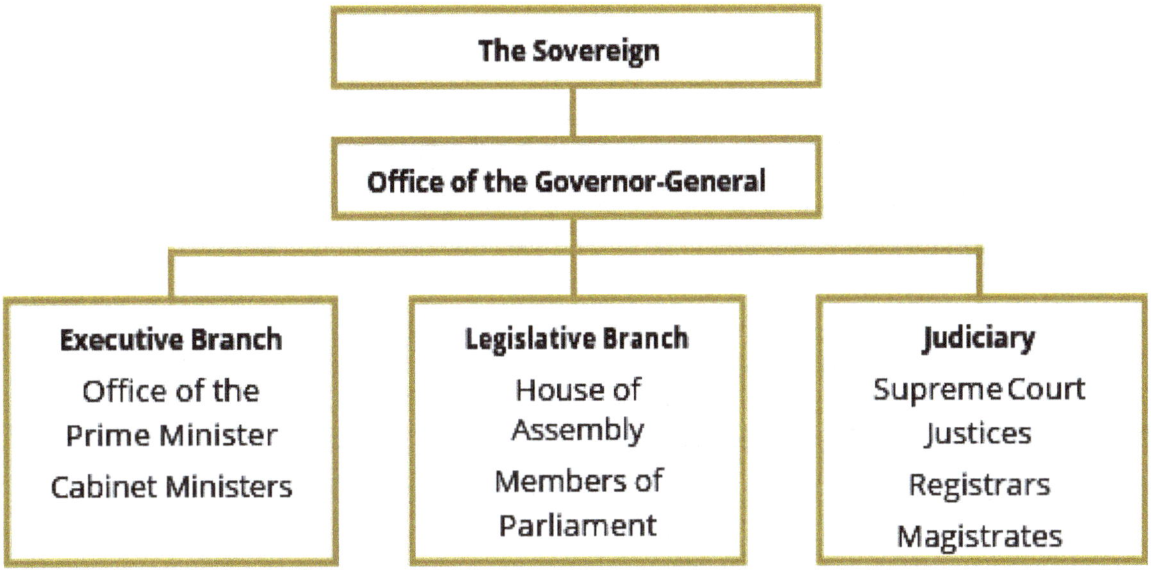

35. Which pair of countries in the Caribbean would have a government structure like that shown above?

A. St. Lucia and Grenada
B. Cuba and Haiti
C. Trinidad and Tobago and Puerto Rico
D. Dominican Republic and Saint Martin

36. Which branch of government would be responsible for the day-to-day administration of a country's affairs?

A. Executive branch
B. Legislative branch
C. Judiciary branch
D. Local branch

37. Paula's father is a Justice of the Peace. With which arm of government does he work?

A. executive branch
B. legislative branch
C. judicial branch
D. local branch

38. Closeness to the sea is a factor that affects all Caribbean countries since they have shores on the Atlantic Ocean. Due to this fact, they enjoy a tropical marine climate.

Marine means __________ .

A. Related to the sea or ocean
B. Related to oxygen or air
C. Related to disasters or storms
D. Related to the environment or the land

39. Which of the following is NOT a result of global warming?

A. Increase in world temperatures
B. Rise in sea levels
C. Tropical storms and hurricanes
D. Increased food supply

40. Paula is the president of the environment club at school. She is collecting bottles to fund the making of make benches. Paula is practising:

A. Recycling
B. Reducing
C. Economising
D. Reforestation

ANSWER SHEET #2

Use the answer sheet to record your answers. Shade in the circle that corresponds to your answer for each test item.

Name: ______________________ **ID no:** ____________ **Age:** ____

School: __________________ **School address:** ______________________

1	A	B	C	D	21	A	B	C	D
2	A	B	C	D	22	A	B	C	D
3	A	B	C	D	23	A	B	C	D
4	A	B	C	D	24	A	B	C	D
5	A	B	C	D	25	A	B	C	D
6	A	B	C	D	26	A	B	C	D
7	A	B	C	D	27	A	B	C	D
8	A	B	C	D	28	A	B	C	D
9	A	B	C	D	29	A	B	C	D
10	A	B	C	D	30	A	B	C	D
11	A	B	C	D	31	A	B	C	D
12	A	B	C	D	32	A	B	C	D
13	A	B	C	D	33	A	B	C	D
14	A	B	C	D	34	A	B	C	D
15	A	B	C	D	35	A	B	C	D
16	A	B	C	D	36	A	B	C	D
17	A	B	C	D	37	A	B	C	D
18	A	B	C	D	38	A	B	C	D
19	A	B	C	D	39	A	B	C	D
20	A	B	C	D	40	A	B	C	D

Score ____________ **out of 40**

This page can be used to write notes.

TEST #2

Answer the questions below by selecting the correct responses.

1. Which of the following ethnic groups cannot be described as voluntary immigrants?

   A. Chinese
   B. Indians
   C. British
   D. Africans

2. Which term best fits the description below?

   'A system by which powerful countries claim weaker countries as their own.'

   A. Colonisation
   B. Emancipation
   C. Immigration
   D. Indentureship

3. My friend, Gina, said that she will be wearing the traditional dress from her country to our heritage day celebration at school. She wore a strip of unstitched cloth wrapped around her waist, with one end draped over her shoulder. During the day, she draped it over her body in various styles.

   Select the ethnic group to which Gina belongs and the name of the traditional dress worn.

   A. Chinese: cheongsam
   B. Japanese: kimono
   C. Jamaican: bandana
   D. East Indian: sari

4. Some of the ___________ went back to their homeland after the period of indentureship, but some stayed and set up grocery shops. To which group does this statement refer?

   A. East Indians
   B. Africans
   C. Chinese
   D. Tainos

5. Which TWO individuals were very influential in the gaining of independence for Haiti?

   A. Marcus Garvey
   B. Napoleon Bonaparte
   C. Toussaint L'ouverture
   D. Jean-Jacques Dessalines

6. Which of these is a distinguishing feature of an independent nation?

   A. Self-government
   B. Overpopulation
   C. Own airline
   D. Wealth

7. Which of the following is NOT an accurate protocol for the use of the national anthem?

   A. Stand with heels together with heads faced forward.
   B. Men remove their hats.
   C. Sing or play at the raising and lowering of the national flag.
   D. Play every morning and evening.

Read the passage below, then answer the questions 8, 9 and 10.

Mr Chang lived in China with his wife and two children. They were very poor and often did not have food to eat or clean water to drink. Mr Chang searched daily to find stable employment but was unsuccessful. One day, while on his usual job hunt, he saw a poster recruiting strong, young men and women to travel to the West Indies for work for a designated time. Mr Chang, seeing no other way to support his family adequately, decided to take advantage of the opportunity.

8. What name was given to the programme being advertised on the poster?

   A. Emancipation
   B. Indentureship
   C. Immigration
   D. Independence

9. Mr. Chang decided that he would take advantage of the opportunity and leave his country to work in the West Indies. Mr Chang would be an/a _________ from China.

   A. Emigrant
   B. Immigrant
   C. Foreigner
   D. Enslaved

10. Which of these were push actors for Mr. Chen?.

   A. Poverty
   B. Famine
   C. Employment
   D. Unemployment

11. On the map below are countries marked with a dot. Which groups originated from these countries?

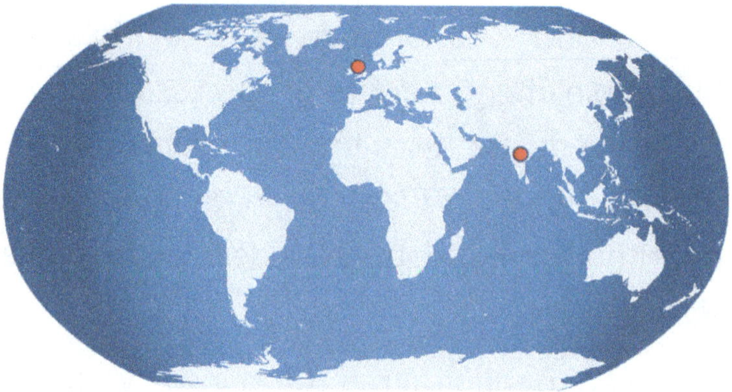

A. Russians and Indians

B. Spanish and Chinese

C. British and Chinese

D. British and Indians

12. The indentured servants brought their own festivals and celebrations to the Caribbean. Which of these are all religious festivals?

A. Divali, Hosay, Eld al-Fitr

B. Crop Over, Chinese New Year, Hosay

C. Eld al-Fitr, Divali, Crop Over

D. Hosay, Chinese New Year, Divali

13. Using information provided in the paragraph below, select the answer that best represents how each country gained its independence.

Slavery lasted for many years in the Caribbean and even after its abolition, the countries remained colonial territories. After years of being colonies, Haiti, Cuba, Jamaica and many others became frustrated by the government of their colonisers. Understandably, they wanted to rule over their own affairs. As a result, brave individuals led the charge and eventually, independence was granted. Cuba sought independence by revolts, with the help of the United States of America. Jamaica, on the other hand, mainly negotiated with Britain for its independence.

A. Jamaica had diplomatic talks with the British government. The Americans helped Cuba with their attacks because Cubans do not speak English.

B. Jamaica had diplomatic talks with the British government. Cuba, aided by the Americans, rebelled.

C. Cuba revolted because Haiti was looking to them and the Americans for support. Jamaica had many revolts and rebellions.

D. All Caribbean countries had a similar history and share a similar path to independence.

14. Select the correct years to show when Jamaica, Cuba and Haiti gained independence.

A. Cuba, 1902; Haiti, 1804; Jamaica, 1962

B. Haiti, 1655; Cuba, 1902; Jamaica, 1944

C. Jamaica, 1834; Cuba, 1804; Haiti, 1902

D. Cuba, 1965; Jamaica, 1962; Haiti 1849

15. Which of these Caribbean countries is a dependent territory?

A. Cuba

B. The Cayman Islands

C. Barbados

D. Jamaica

16. Examine the profile below and identify the hero to which it refers.

- *National Hero of Jamaica*
- *Was a Jamaican statesman*
- *A Rhodes Scholar*
- *Became one of Jamaica's leading lawyers in the 1920s*
- *An advocate of universal suffrage, which was granted by the British colonial government to the colony in 1944*
- *Born: July 4, 1893, Manchester*
- *Died: September 2, 1969, Kingston*
- *Spouse: Edna (m. 1921)*
- *Place of burial: Jamaica*
- *Mother: Margaret Shearer*
- *Children: Michael, Douglas*

A. Marcus Garvey
B. Norman Manley
C. Alexander Bustamante
D. Samuel Sharpe

17. Identify the mountain range labelled with the blue triangle.

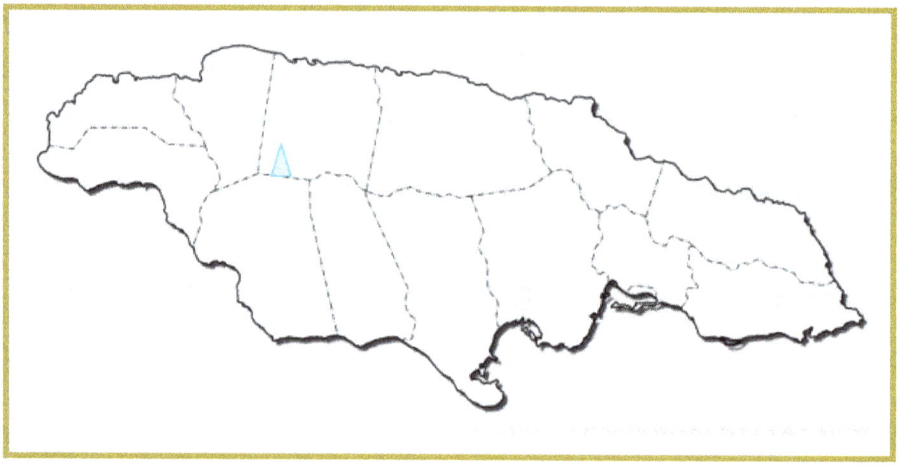

A. Blue Mountain
B. Dry Harbour Mountain
C. Cockpit Country
D. John Crow Mountain

Read the passage below then complete questions 18 and 19.

In the small farming community of Wilson Peak, the community members use the hillsides and mountain slopes to settle and grow crops, such as coffee and banana. They also cut trees for burning charcoal to earn extra income. They believe that growing crops on the slopes results in a good yield when it is time to harvest.

The farmers are now complaining that their main source of income is being threatened and are asking for help from the National Environment and Planning Agency (NEPA) and the Ministry of Agriculture. Representatives from both entities met with the community members in a town hall meeting to discuss possible preventative and intervention measures. The farmers are now happy that they may be able to save their crops.

18. Using the information from the passage, choose the answer that lists three ways that the small farming community of Wilson Peak has impacted the environment.
    A. Clearing the land, planting and reaping crops, mining the soil
    B. Meeting with NEPA, planting and reaping crops, clearing the land
    C. Burning coal, planting and reaping crops, settling on land
    D. Burning coal, mining the soil, meeting with NEPA

19. Identify the farmers' main problem.
    A. Lack of political support
    B. adequate access to fertilizers
    C. More buyers for their crop
    D. Soil erosion

20. Which of these statements is false?
    A. Lines of latitude and longitude are imaginary.
    B. Lines of latitude intersect at 90°.
    C. The Equator is a line of latitude.
    D. Lines of longitude run north to south.

21. You have been asked to label the lines indicated on the globe below. Choose the option which correctly labels the lines from north to south.

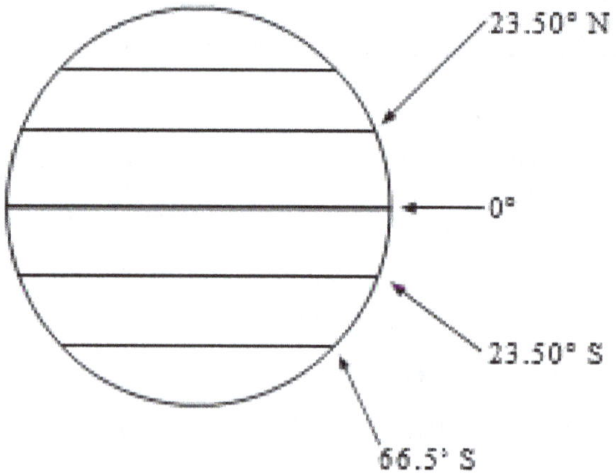

A. Tropic of Capricorn, Equator, Tropic of Cancer, Arctic Circle
B. Tropic of Cancer, Equator, Tropic of Capricorn, Arctic Circle
C. Antarctic Circle, Equator, Tropic of Capricorn, Arctic Circle
D. Tropic of Cancer, Equator, Tropic of Capricorn, Antarctic Circle

22. What line of latitude marks the hottest section of the Earth?
A. Equator
B. Tropic of Capricorn
C. Antarctic Circle
D. Tropic of Cancer

23. I was taken from my homeland and brought to this country with many other dark-skinned people. Some of us died along the way from sickness or at the hands of our white captors. I am now forced to work in cane fields or be killed.

To which group does this person belong?

A. English
B. Chinese
C. Africans
D. East Indians

24. When handling the flag of Jamaica, one protocol states that it should never touch the ground. Why do you think this protocol exists?

    A. So it does not get dirty
    B. To show respect for the national emblem
    C. To prevent short persons from touching it
    D. So that the emblem can be hoisted more easily

25. The wood of the Lignum Vitae tree is used as a propeller shaft bearing in ships.

    Why do you think this is so?

    A. Its appearance makes it very attractive.
    B. No other wood is available to be used.
    C. It is very durable.
    D. It is the national tree.

Read the dialogue between the two students below and then respond to questions 26 to 28.

Jessica: Carl, in class I learned about the main river on each continent. I cannot seem to remember the correct order in length and the continents on which the Darling, Amazon, Volga and Nile rivers are found.

Also, is there a river on Antarctica?

Carl: Yes, Jessica. I remember the correct order and the continents on which each is found. The Nile is in Africa and is the longest. The Amazon River is located in South America. It is the second longest river in the world. The Darling River in Australia would be next and then Volga River in Europe.

There is no river in Antarctica.

26. Which of Carl's statements about the main rivers in the world is incorrect?

A. The Nile is the longest river.
B. Volga River is in Europe.
C. There is no river in Antarctica.
D. The Amazon river is located in South America.

27. Which of the statements below can replace the incorrect statement?

A. There is a river in Antarctica called the Onyx River.
B. The Amazon is not a river but a forest.
C. Darling River is in Europe.
D. There is no river in Africa.

28. During the students' conversation, they failed to identify two other rivers and their continents. Which options shows rivers correctly matched to their continents?

A. Yangzte in Asia
B. Mississippi in North America
C. Mississippi in Asia
D. Yangzte in North America

29. Craig created the coat of arms above for his project on national symbols. He needs to label the part marked with an arrow. Which option should he choose?

A. Mantling
B. Crest
C. Bearing
D. Helmet

30. Select the option which shows the two statements that are true.

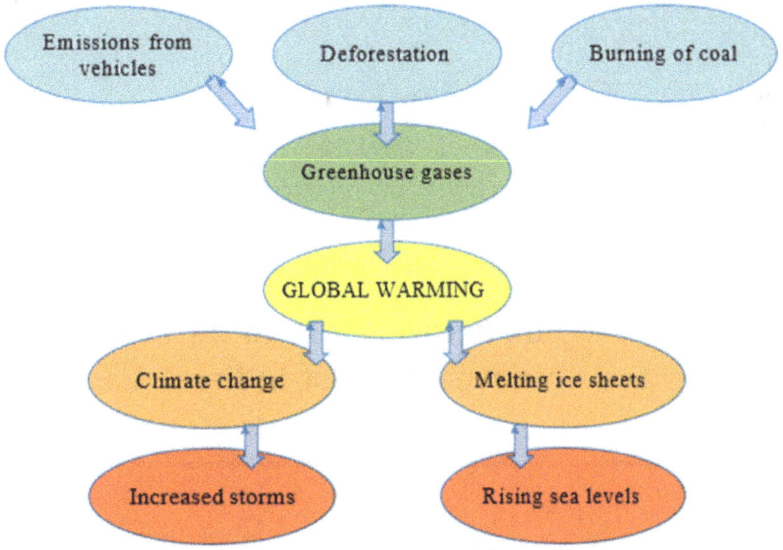

I. Global warming is caused only by deforestation.
II. When the atmosphere becomes polluted by vehicle emissions, greenhouse gases are released.
III. The rising sea levels cause global warming.
IV. Climate change is a good thing.
V. When the ice sheets melt, we have more water to drink and play in.
VI. One effect of global warming is more frequent hurricanes.

A. i and iii
B. ii and vi
C. v and iv
D. iii and v

31. Amanda is a citizen in Jamaica and Canada. She does not wish to give up her citizenship in either country.

Which term best describes the situation outlined above?

A. Dual citizenship
B. Membership
C. Indentureship
D. Relationship

32. Jamaica is a democratic country. This means that the government is chosen by the ______________.

A. Citizens
B. Monarch
C. The Governor General
D. The Prime Minister

The characteristics in the box below relate to different festivals brought to the Caribbean by the Chinese and East Indians. Use them to respond to items 33 and 34.

I. Dragon dance
II. Celebrated by Muslims in Guyana
III. Marks the end of the sugar cane harvesting season in Barbados

33. Which of the festivals below would include the dragon dance?

A. Hosay
B. Easter
C. Crop Over
D. Chinese New Year

34. Which pair of characteristics would be associated with religious festivals?

A. I and III
B. II and IV
C. I and II
D. IV and III

35. Citizenship means more than simply being born somewhere. It also includes active participation as a member of society. Which of the following is NOT a responsibility of a citizen?

A. Paying all taxes
B. Protecting the environment
C. Protecting your neighbour's property
D. Becoming an educated and informed member of society

36. Jada's father is disgruntled with the ruling of the judge at the Supreme Court. What is the name of the next court that his case could be heard?

    A. Petty Sessions Court
    B. The Resident Magistrates/Parish Court
    C. The Court of Appeal
    D. The Circuit Court

37. Which statement is NOT true about local government?

    A. Every community has a mayor.
    B. Local government provides goods and services which are paid for by our taxes.
    C. Each parish council/municipality is led by a mayor.
    D. Every mayor is a councillor, but not all councillors are mayors.

38. Patricia and Paula were at home playing. Detective Taylor and Constable Williams came over and searched their house for no stated reason. Which right of the citizen did the officers violate?

    A. Right to life
    B. Right to education
    C. Protection from inhumane treatment
    D. Right to privacy

39. Which set of words BEST describes the government service depicted in each picture?

A. Heart, water, energy, security, school
B. Health, ocean, energy, police, education
C. Heart, water, energy, military, school
D. Health, water, energy, security, education

40. Which country in the Caribbean would be MOST AFFECTED by a rise in the sea level?

A. Jamaica
B. Grenada
C. The Bahamas
D. Cuba

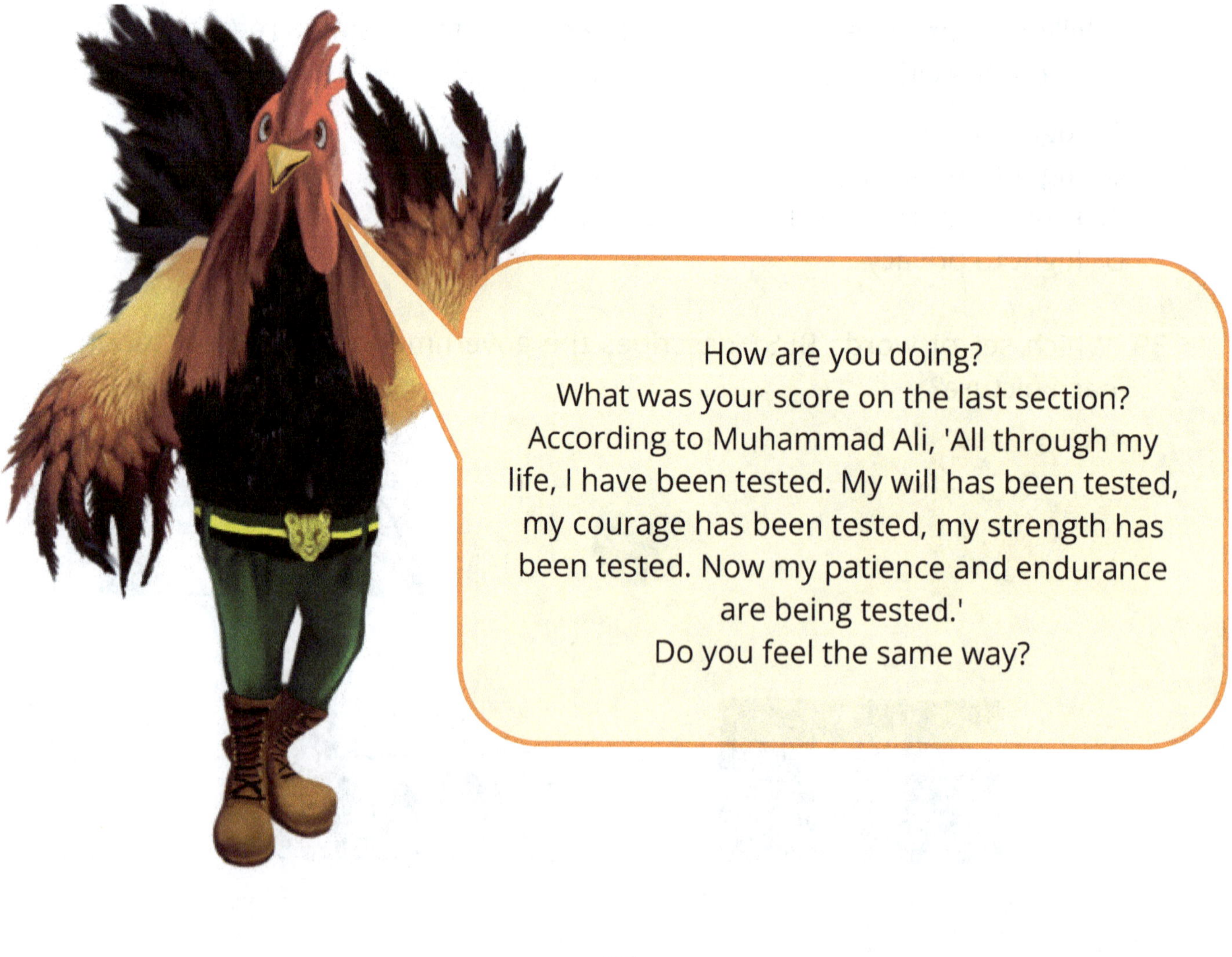

How are you doing?
What was your score on the last section?
According to Muhammad Ali, 'All through my life, I have been tested. My will has been tested, my courage has been tested, my strength has been tested. Now my patience and endurance are being tested.'
Do you feel the same way?

ANSWER SHEET #3

Use the answer sheet to record your answers. Shade in the circle that corresponds to your answer for each test item.

Name: ______________________________ **ID no:** ________________ **Age:** ____

School: __________________________ **School address:** __________________________

1	A	B	C	D
2	A	B	C	D
3	A	B	C	D
4	A	B	C	D
5	A	B	C	D
6	A	B	C	D
7	A	B	C	D
8	A	B	C	D
9	A	B	C	D
10	A	B	C	D
11	A	B	C	D
12	A	B	C	D
13	A	B	C	D
14	A	B	C	D
15	A	B	C	D
16	A	B	C	D
17	A	B	C	D
18	A	B	C	D
19	A	B	C	D
20	A	B	C	D
21	A	B	C	D
22	A	B	C	D
23	A	B	C	D
24	A	B	C	D
25	A	B	C	D
26	A	B	C	D
27	A	B	C	D
28	A	B	C	D
29	A	B	C	D
30	A	B	C	D
31	A	B	C	D
32	A	B	C	D
33	A	B	C	D
34	A	B	C	D
35	A	B	C	D
36	A	B	C	D
37	A	B	C	D
38	A	B	C	D
39	A	B	C	D
40	A	B	C	D

Score ______________ **out of 40**

This page can be used to write notes.

TEST #3

Choose the BEST answer for these questions.

Evaluate the social status of each of the following persons and respond to items 1 and 2 below:

A. Enslaved
B. Indentured servant
C. Adult male citizen after 1962
D. Every citizen after 1962

1. Who had the MOST rights and freedom? __________

2. Who had NO personal rights or freedom? __________

3. Interpret the meaning of Jamaica's national motto.
   A. Jamaicans are a united people who come from various backgrounds.
   B. We are all different but equal as Jamaican citizens.
   C. One leader speaks for many people from many places.
   D. All Jamaicans follow the same leader.

4. Which of the following historical figures was an Italian explorer most well-known for exploring the New World and who arrived in Jamaica in 1494?
   A. King George
   B. Bob Marley
   C. Prince Albert
   D. Christopher Columbus

5. As a project for your history class, you must create a brochure advertising Kingston. Which of the following topics should you include in your brochure about Kingston?
   A. Geographical information
   B. Economic information
   C. Historical information
   D. All of the above

6. Which answer accurately completes the blank timeline?

   1655 –

   1807 – The ending of the slave trade

   1832 - Sam Sharpe was hanged

   1838 –

   1865 – Morant Bay Rebellion

   A. The British captured Jamaica in 1655. All adult males were given the right to vote in 1838.
   B. The British captured Jamaica in 1655. All slaves were emancipated in 1838.
   C. Jamaica became a Spanish colony in 1655. In 1838, all children six years and younger were given full freedom.
   D. Jamaica gained independence from Spain in 1655. Universal adult suffrage was given in 1838.

7. 'Spiritual victory of light over darkness, good over evil and knowledge over ignorance.' This is a good description of:

   A. Diwali
   B. New Year's Day
   C. Crop Over
   D. Hosay

8. Which source would be the BEST reference to write a report about Marcus Garvey?

   A. Encyclopedia article
   B. History textbook
   C. His personal diary
   D. A candid picture of him

9. Which of the following is OUT of chronological order?

   A. Emancipation, universal adult suffrage, independence
   B. British capture of Jamaica, end of the slave trade, emancipation
   C. Sam Sharpe Rebellion, emancipation, independence
   D. Universal adult suffrage, emancipation, independence

10. Why is the hummingbird important to Jamaican citizens?

    A. It is an essential part of the Jamaican economy.
    B. It is a beautiful bird, loved by all Jamaicans.
    C. It is indigenous to the tropical environment.
    D. It is a chosen national symbol of Jamaica.

11. Which of the following is a TRUE statement?

    A. Higher elevations have warmer temperatures than lower elevations.
    B. Lower elevations have more rain than higher elevations.
    C. Neither
    D. Both

12. Which of the following is a landmass that DOES NOT have at least three sides surrounded by an ocean or sea?

    A. Gulf
    B. Peninsula
    C. Island
    D. Isthmus

13. Which of the following is a FALSE statement?

    A. You must be born in Jamaica in order to be a Jamaican citizen.
    B. Jamaicans have complete personal freedom to do whatever they choose.
    C. Neither
    D. Both

Read the statement below and then answer question 14.

'With hotter temperatures, coastal encroachment, shorter and more intense periods of rain, longer and more intense periods of drought and more frequent and destructive hurricanes, climate change is our daily reality.'

The Most Honourable Andrew Holness, ON, PC, MP

14. Which of the following is NOT an effect of climate change?

A. Increased temperature
B. Shorter periods of drought
C. Increased hurricanes
D. Loss of our shores

15. Which of the following are rights of Jamaican citizens?

A. Employment and income
B. Privacy and protection
C. Neither
D. Both

16. Examine the timeline below. Which event would best fit in the blank timeline?

1494 – Christopher Columbus landed in Jamaica and named it the West Indies.

1509 – First Spanish colonists settled in the St. Ann's Bay area.

1655 –

1673 – Henry Morgan was selected to be the Lieutenant Governor of Jamaica.

1739 – Maroon wars ended after years of fighting.

A. Jamaicans were allowed to vote for the first time.
B. The Jamaican government was formed.
C. The English arrived in Jamaica and attacked the Spanish.
D. The Maroon wars began.

17. Which island is independent?

A. Cayman Islands
B. Dominican Republic
C. British Virgin Islands
D. Montserrat

18. Why are the pineapples included on the shield of Jamaica's coat of arms?

A. To show Jamaica's favourite fruit
B. To signify friendship
C. As a symbol of Jamaica's native fruits
D. To include colourful symbols

19. Dwayne's job at school is to take the flag down from the flagpole. Then he is supposed to fold it neatly and store it until it is ready to be used again.

Identify TWO things which Dwayne should not do with the flag?

A. Let it hit the floor
B. Put it in a drawer
C. Let it get dirty
D. Fold it

In this 'odd man out' exercise, circle the answer that does NOT fit with the others for question 20.

20. Which does NOT belong?

A. Legislative
B. Executive
C. President
D. Judicial

21. What is the definition of a bay?

A. A piece of land that is surrounded by water
B. A large body of water that is hundreds of miles in area
C. A stream of water that meanders through the country and empties into a larger body of water
D. A small body of water that juts inward toward the island and empties into a larger body of water

Examine the table below about the average size of the continents and oceans.

Continent or Ocean	Size in square km
continents	150
oceans	362

22. What is the ratio of oceans to continents?

A. 1 to 2.4　　B. 2.4 to 1　　C. 2 to 1　　D. 5 to 1

23. Which of the following characteristics refers to both lines of latitude and longitude?

A. Runs east to west/west to east
B. Runs north to south/south to north
C. Divides the earth into climatic zones
D. Is used to locate places

24. Paul wants to convince the government to have a 'Jamaica Endangered Species Day'. This is to recognise Jamaican plants and animals that may go extinct if we do not protect their habitats.

    Which statement CANNOT be used to convince the government?

    A. Laws can convince the Jamaican people to protect endangered species.
    B. The Jamaican national bird is an endangered species.
    C. The endangered plants and animals play no vital roles in the ecosystem.
    D. We do not fully know the medicinal value of some of the plants that may go extinct.

25. I visited the Blue Mountains Range and collected a soil sample. Which option would best describe the main characteristics of the soil sample?

    A. Light brown soil.
    B. Very sparse and rocky soil
    C. Dark, clumpy, rich fertile soil
    D. Light and dusty red soil

26. Which THREE of the following are opinions?

    A. Jamaica's beaches are the best, so tourists love to visit.
    B. Jamaica has fifteen interesting festivals each year.
    C. Jamaica's average temperature is in the 70s.
    D. Jamaica's mountains are popular recreational spots.

27. Which of the following statements is true about culture?

    A. Culture is the qualities you are born with.
    B. Culture includes traditions and beliefs.
    C. Culture is based on people's physical features.
    D. Culture does not change over time.

28. Which TWO of the following are justifiable facts?

A. Alexander Bustamante formed the Bustamante Industrial Trade Union and founded the Jamaica Labour Party.
B. Alexander Bustamante was born on February 6, 1884 in Manchester.
C. Alexander Bustamante was the only living national hero to receive this honour.
D. Alexander Bustamante's first job was as an accountant.

29. Which TWO of the following are branches of the Jamaican government?

A. Judicial
B. Ancestry
C. Population
D. Legislative

30. Which THREE of the following are good team-building skills?

A. Compromise
B. Gossip
C. Honest communication
D. Respectful disagreement

31. What are TWO ways for a person to become a Jamaican citizen?

A. Being born in Jamaica
B. Working in Jamaica
C. Being born to Jamaican parent/parents
D. Owning property in Jamaica

32. Which TWO of the following mountains are MOST likely to attract tourists to Jamaica?

A. Blue Mountain Peak
B. Mount Vesuvius
C. Mount Everest
D. John Crow Mountain

33. Which TWO of the following would MOST likely be enjoyed in a tropical climate?
    A. Snow skiing
    B. Snorkeling
    C. Rafting
    D. Ice hockey

34. Which TWO of the following are parts of a river?
    A. Body B. Source C. Texture D. Mouth

35. Which TWO of the following are myths or misconceptions about atmospheric pollution?
    A. Atmospheric pollution only occurs in very highly populated areas.
    B. Atmospheric pollution only affects people with allergies.
    C. Atmospheric pollution affects everyone, but the very young and the very old are particularly susceptible.
    D. Everyone should try to help stop atmospheric pollution.

36. Identify a group which is made up of only a set of one of the following: land masses, landforms, water bodies or coastal features.
    A. Cliffs, beach, peninsula, isthmus
    B. Gulf, hill, mountain, sea
    C. Ocean, plain, lake, river
    D. Mountain, hill, plateau, lake

37. It is most likely to rains or snows on the ______________ side of a mountain.
    A. Windward
    B. Highland
    C. Leeward
    D. Colder

38. Read the information and answer the question below.

While working in a group at school, Kevin and Tina disagree about which products or goods are important to Jamaica's economy. How should the group resolve this conflict amicably so they can complete their task.

A. The rest of the group should decide without Kevin or Tina giving their input.
B. Kevin and Tina should argue until one of them gives up.
C. Kevin and Tina should discuss their ideas and negotiate.
D. The group leader should choose the best idea.

39. Your teacher wants you to use multiple sources to research universal adult suffrage and voting in Jamaica. You find a useful source that explains the history of Jamaicans gaining the right to vote. How should you select the other sources you should gather based on your research topic?

A. Use sources that are referred to you by your friends.
B. Find other reliable sources that pertain to the history of Jamaican voting.
C. Find all the websites related to Jamaican voting.
D. Find the sources about voting around the world.

40. Where is Jamaica's motto officially written?

A. On the coat of arms
B. On official letterheads
C. On the flag
D. In the constitution

CHEETAH SOCIAL STUDIES

Unit Assessment and 40-Questions Tests Answers and Explanations

TERM 1: UNIT 1

Theme: How Can We Promote and Preserve our Caribbean Culture?

1. Matching

1. indentured servant-H; indentureship-I; contract-B; festival-C; carnival-A; immigrant-D; migration-F; push factors-E; pull factors-G

2. B: push factor(s), pull factor, contract, indentureship, indentured servant (s), secular and religious

3. A: Heritage describes practices or traditions passed down to others. Culture describes the shared beliefs, values and social norms of a group. An ethnic group is made up of people who share physical and behavioural characteristics and language.

4. C - heritage

5A. Africans - 1513
Jews - 1494
Chinese - 1854
Indians - 1845
German - 1834

5B.

The Jews first came to the island during the Spanish inquisition in 1494 from Portugal and Spain. The first set of Africans arrived in 1513 as enslaved to the Spaniards. The Germans came in 1834 as indentured labourers. The Chinese came to Jamaica in 1854, after the East Indians who came over in 1845 as indentured servants.

6. C: There are 360 years between the Jews migrating in 1494 and the Chinese coming to Jamaica in 1849.

7 C: Y is located in China and X is located in India

8. D: Z represents the Germans since it marks Germany on the map.

9.

Condition or event	Push factor	Pull factor
A. restricted freedom	√	
B. promise of jobs		√
C. Spanish Inquisition	√	
D. religious oppression	√	

10. D: Both the Chinese and East Indians migrated to the Caribbean as contract workers for better job opportunities and life, remaining even after the indenture- ship had ended.

11. Answers may vary.

12. B: The Chinese and East Indians were provided with some things, including clothing and tools. They worked for long hours and were not allowed to leave the plantation without permission, or they could be fined. However, they were not provided with regular medical check-ups, as was promised by the plantation owners.

13A. Answers may vary.

B. Answers may vary.

C. A) does not support B) supports C) supports D) does not support

14. D: Immigrants likely supported and stayed within their ethnic groups.

15. D: Chinese grocers impact the expanding food and beverage industry on the island.

16. C: The Chinese and East Indians were successful in businesses that involved retail, wholesale and manufacturing. They were skilled in many areas allowing them to start small family businesses, which later developed into bigger businesses requiring additional workers.

17A. sugar - goods rum - goods bauxite - goods music - service sugar - goods

17B. C: Jamaica's economy relies on both selling goods and offering services. Services offered in the tourism and music industries and goods, such as produce and food products, help build the economy.

18. C: Sugar has had the greatest impact on the Jamaican economy. Sugar plantations date back to the island's colonial days.

19. B: Migrants to a new country keep traditions of their culture by bringing their traditions, celebrations and religious ideas to the new country. They share these traditions with the people located in the new country.

20. D: These activities commemorate the death of Christ and not His birth.

21. Answers may vary.

22. A.

Statement	Preserves	Does not preserve
A. We can preserve our Caribbean culture by learning about and celebrating our food, religion, performing arts, sports and language.	preserves	The East Indians brought festivals and celebrations.
B. We can preserve our Caribbean culture by limiting our exposure to other cultures.		does not preserve (Exposure to other cultures is useful and enriching. It does not necessarily threaten our own culture.)
C. We can protect our landmarks, heritage sites and important buildings by not destroying or defacing them.	preserves	
D. We can preserve our culture by insisting that emigrants continue to share aspects of our culture with others in their new home: our food, music, dance and language.		does not preserve (We cannot force people to share their culture.)

22. B. Answers may vary.

22. C. Answers may vary.

23. B and D: Promoting Caribbean culture can be done in various ways. Passing down information from one generation to the next through storytelling is one such way. Creating spaces dedicated to cultural and creative performances is another.

24. A and D: Learning about Caribbean culture helps increase the nationalistic idea of the people. It also allows people to learn about their history.

25. B and D: Cultural experiences and events promote Caribbean culture, as does having more regional activities.

26. product: clothing, service: dressmaking/ tailoring

Explanations will vary.

27.

Statement	Cultural value
A. The East Indians brought their festivals and celebrations like Divali and Hosay to the Caribbean.	The East Indians brought festivals and celebrations.
B. Chinese developed their businesses until the small grocery shops grew into large enterprises. They embraced not only retailing, but also wholesaling and other types of activities.	no cultural value
C. The Indians brought their religions such as Hinduism and Islam, which are still practised.	no cultural value
D. The Chinese represents a very small proportion of the Jamaican population.	
E. The Chinese brought some of their food preparation techniques to the Caribbean.	food preparation techniques add cultural value is missing

28. Answers may vary.

29. C and D: The government, through the Jamaican National Heritage Trust, is investing money and making laws to protect historical places throughout the island. Unfortunately, individuals continue to deface and destroy our heritage.

30. C: Source 3. This source describes a speech about people in Jamaica defacing heritage sites.

31. C: Using different aspects of the Jamaican culture in cultural and creative industries brings investments, money and provides jobs for people. Through these economic industries, money is reinvested into areas around Jamaica.

32. Answers will vary.

How are you doing? Have you reviewed all your responses to the questions? Do you know why you got any of the questions wrong?
Come wid mi! Mek wi continue.

TERM 1: UNIT 2

Theme: Our Common Heritage

1. Answers will vary.

1B. Glossary

commonwealth - an international association of nations symbolically headed by the British monarch

franchise - an individual or a group that is authorised to be an agent of a company's products or services

political party - a group of people with shared beliefs and ideas about a government's function

revolution - an overthrow of a government in favour of a new one

trade union - an organised association that works to negotiate for and protect workers' rights

universal adult suffrage - the right of all adult citizens to vote without discrimination

2. **D.** Fill in the Blank: revolution, independence, colonial rule, coloniser

3. **A.** monarchy, nation, constitution, commonwealth, political party suffrage

4. **C.** self-government, franchise(s), trade union

5A. **C.** 1962 Jamaica gained independence

5B. **D:** The Spanish colonists settled in 1509.

6. **A.** 161 years

7. **D:** Sources 1, 2 and 3 support the statement.

8A. **B**: All three national heroes believed that the people would benefit from a self-governing and independent Jamaica.

8B. Answers may vary. Answers may include the following:

- It has factual information which can be verified.
- There are no opinions.
- The tone is objective.

9. Answers may vary. Answers may include:

- exercising franchise or the right to vote
- appreciating those who fought for independence.

10. Answers may vary. Answers may include:

- valuing employers

- honouring minimum wage
- examining what happens when people are overworked

11. C: Colonies had revolutions from their mother countries because they desired to be free, have self-rule and create self-governments in the nations.

12. D. Republic of Cuba, Jamaica, Haiti

13. C: Marcus Garvey went to prison in the USA.

14. Answers may vary.

15. A, B, C; Haiti, Cuba and Jamaica commemorate their independence by having festivals with food, drinks, parades, dance and music.

16. Answers may vary.

17. Answers may vary. Answers may include the flag-raising ceremony, presentation about persons who made it possible, national anthem, Grand Gala, national costumes and parades.

18A. Answers may vary.

18B. Answers may vary.

19.

Caribbean nation	Dependent	Independent
Anguilla	√	
Barbados		√
Jamaica		√
The Bahamas		√
British Virgin Islands	√	

20. A. The concept of self-government allows the citizens of a country to participate in the government by serving as elected officials or participating in elections.

21. C. It is important to make persons aware of the contributions made towards independence.

22. Answers will vary.

23. Answers will vary.

24. C. There were more sources of revenue, including mining and tourism.

25. C. low wages

26. B. Jamaica became resourceful and self-sufficient.

27. Answers may vary.

28. Students will have varied posters

29.

Author	Title	Publisher	Date of Publication
Dennis, Hope E.	E. We Want Justice! Jamaica's Journey to Nationhood	Jamdown Publishers	E. 2009
Munroe, Garth B.	B. Slavery to Independence: A Caribbean Perspective	B. Begonia Publishers	B. 1965
Saddler, Barbara A.	A. Pre-independence History of Jamaica	A. Beautiful Brooks	A. 1976

30. Answers will vary.

'We have become not a melting pot but a beautiful mosaic. Different people, different beliefs, different yearnings, different hopes, different dreams.'
— Jimmy Carter
What do you think? Come wid mi! Mek wi continue.

TERM 1: UNIT 3

Theme: Living Together

1. Emblem-F; Flag-D; bearing-C; symbol-A; nationhood-H; Crest-B; patriotism-E;

1B. a) motto - an expression of beliefs or values

b) anthem - a song with special meaning for people, a place, organisation or country

c) coat of arms - a design with a shield, crest and arms that has meaning

2. Fill-in-the-blanks

nationhood, emblems, anthem, motto, flag, patriotism, coat of arms

3a. D: Flag, anthem and flower are important symbols of a nation upon becoming a new, independent nation.

3b. Answers may vary.

3c. Answers may vary.

3d. Answers may vary.

4. C

5. C

6a. Answers may vary.

6b. Answers may vary.

7. D

8. Answers may vary.

9. A. Black shows Jamaican's creativity and strength; gold represents the island's richness and sunshine.

10. B. Protect Jamaicans from those who may want to harm us.

11. B

12. Answers may vary.

13. D: Food for the Poor ensures that men, women and children receive medical, financial, social or spiritual help without discriminating against race, status or creed.

14. Answers may vary.

15. A-C. Answers will vary.

16. A-B. Answers will vary.

17. Answers may vary.

18. B.

19. C. The Governor General should always be referred to as Your Excellency.

20. A. Playing the national anthem is the proper protocol to announce the Prime Minister's arrival.

21a. Answers will vary.

21b. Answers will vary.

TERM 2: UNIT 1

Theme: The Physical Environment and Its Impact on Human Activities

1a. landforms, mountains
landforms, summit, mountains, mountain range, hills, mountains, hills, mountains, slopes, valleys, plateaus, plains, forest reserves

1b.

Physical feature	Tick (√) if found in Jamaica
deserts	
valleys	√
glacier	
coastal plains	√
mountain ranges	√
tundras	
hills	√

1c. 1. I; 2. D; 3. A; 4. J; 5. F; 6. C; 7. H; 8. G; 9. E; 10. b

2. **C.**

3. **D.** landforms, plains, mountain ranges, summit, valleys

4. **A.** mountain, hill, slopes, plateau, forest reserves

5a.

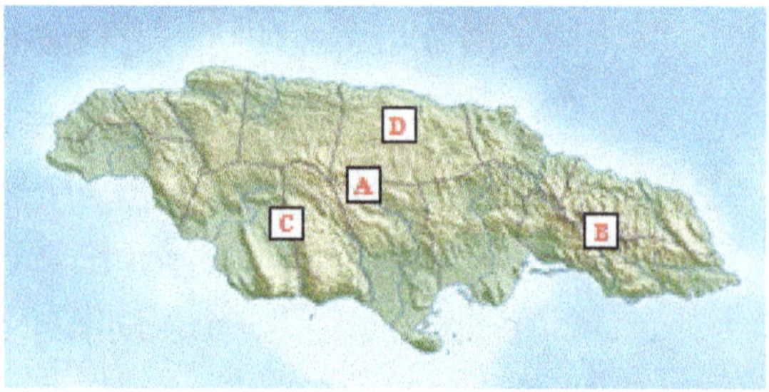

5b. **D** Blue Mountains are to the east.

6a.

Name of mountain	Parish location	Rank (1=lowest 5=tallest)
Gossomer Peak	St. Thomas	3 (1012 m)
Dolphin Head	Hanover	2 (380 m)
Blue Mountain Peak	Portland/St. Thomas	5 (2257 m)
High Peak	Portland	4 (1800 m)
Frasers Mountain	Hanover	1 (15 m)

Source: https://www.britannica.com/place/Jamaica

6b.

Western	Central	Eastern
Dolphin Head	Don Figuerero Mountains	Blue Mountains
Orange Hill	Dry Harbor Mountains	John Crow Mountains
Santa Cruz Mountains	Mocho Mountains	

7. **C.** The Blue Mountain Range contains the highest summit on the island, the Blue Mountain peak.

8. **B.** You would expect to find more rain on the windward side of the mountain because as the air mass rises, the moisture in the clouds is released because colder air cannot hold moisture.

9. Answers may vary. Altitude (or elevation) affects the climate of an area. The higher up you go in altitude, the colder the climate. Coastal regions in Jamaica are low elevation and would be warmer than mountain ranges over 2000 metres above sea level.

10. Answers may vary.

11A-C. Answers may vary.

12 Answers may vary.

13. Answers may vary.

14. A. Andes - South America

15. B. Asia

16 B. By recycling the water bottle, Rashona shows responsibility and care for the environment.

17. A and C. To help the native bird survive, it is important to keep its habitat rich. Passing a law preventing deforestation would help keep the natural resources available for the population of the lizard cuckoo to flourish. Additionally, creating conservation groups would ensure people continuously look out for the bird's protection.

18. C. The best ways to prevent land slippage (landslides) are to create terracing, or contours, plant deep-rooted trees and plant cover crops. All these prevent slippage. A retaining wall will only hold back the eroded soil from sliding to lower surfaces.

19.

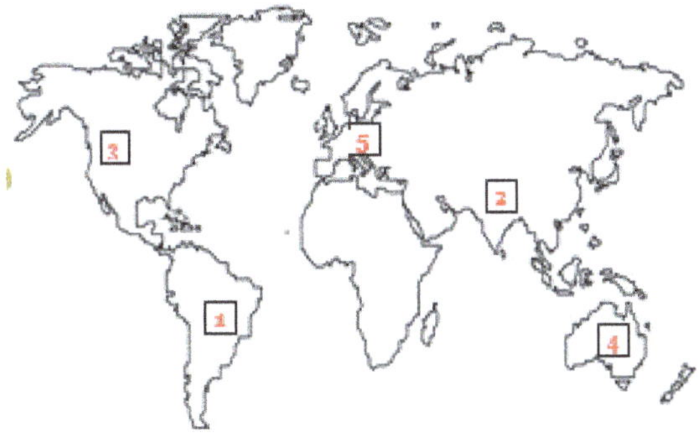

Himalayas - Asia

Rockies - North America

Great Diving Ranges - Australia

Alps - Europe

20. Some of the effects of deforestation include soil erosion, droughts, loss of biodiversity and climate change. Building roads to remove the soil before erosion occurs or building a rock wall or retaining wall are not effects of deforestation.

TERM 2: UNIT 2

Theme: The Physical Environment and Its Impact on Human Activities

1.

Words	Definitions
Example: lake	*a body of water that is inland and completely surrounded by land*
continent	the largest landmass on Earth
island	a small piece of land surrounded by water
ocean	a large body of salt water; the largest body of water on Earth
bay	the smallest inlet of water protected by land
peninsula	a piece of land surrounded by water on three sides

2. Answers may vary.

3. **A.** ocean, sea, lake, river, bays, gulfs

4. **D.** The islands of Jamaica, Hawaii and Japan are all archipelagos. Each is made up of groups of islands.

5. **A.** grid, longitude, latitude, hemisphere, great circle

6. A. Lines of latitude measure degrees north and south of the Equator.

7.

Name of continent	Rank (1=largest 7 = smallest)
Asia	1
Europe	6
Antarctica	5
North America	3
South America	4
Africa	2
Australia	7

Body of water	Rank (1=largest 3=smallest)
bay	3
ocean	1
sea	2

8.

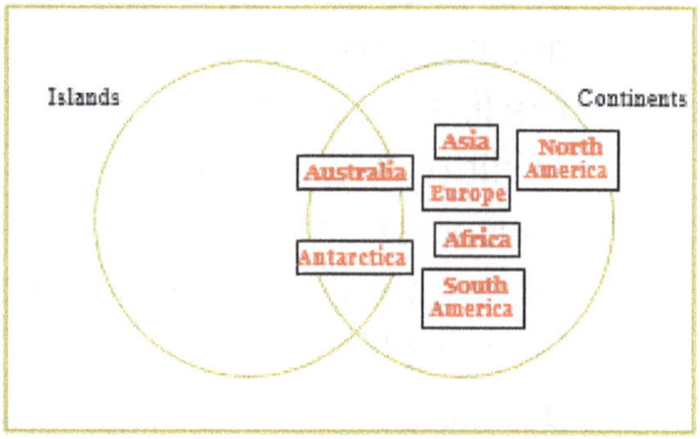

Australia and Antarctica are continents and also islands because they are both surrounded by water.

8B. Answers may vary.

8C.

Waterbody	Collects water	Moves Water	Surrounded by land	Has no boundaries	Freshwater
ocean	X			X	
seas	X			X	
lakes	X		X		X
river		X	X		X
bay	X				

9. Answers may vary.

10. Answers may vary.

11. A

12. D. A sea is a larger body of water. A bay is a smaller inlet of water that is surrounded and protected by land.

13. C. Both seas and rivers are bodies of water.

14.

Landmass	Description of special features
isthmus	a piece of land that connects two larger pieces of land
peninsula	a piece of land that sticks out into the sea form a larger landmass.
continent	one of 7 large landmasses on the Earth
island	an area of land that is surrounded by water

15. absolute, relative location

16. D.

17. C. The place where the Tropic of Cancer and Prime Meridian are located is Africa.

18.

Statement	Tick (√) if accurate statement
Most landmasses are located south of the Equator.	
The Equator is located between the tropic of Cancer and Tropic of Capricorn.	√
Jamaica is located between the Equator and the Tropic of Cancer.	√
There are no landmasses located south of the Tropic of Capricorn.	

19. D. 18°00′N 77°00′ W.

20.

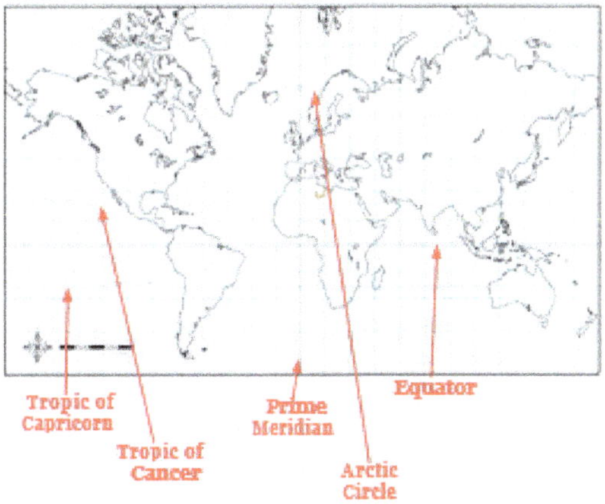

21A. B. Lines of latitude and longitude help make up the global grid system. The lines of latitude measure distance north and south of the Equator. Lines of longitude measure distance east and west of the Prime Meridian.

21B. B. Lines of longitude run north and south from pole to pole and they measure distance east and west of the Prime Meridian. You would not use lines of longitude to measure north or south of the Equator.

22. C. Lines of latitude have 90 degrees north and 90 degrees south.

23 Yes, a location with a lower latitude number means the location is closer to the Equator and has a warmer climate. The higher latitude number means the location is closer to the Polar Regions.

24.

Characteristic	Parallel	Meridians
runs	east to west	north to south
measures	north or south	east or west
number	180	360
measure of 0 degrees	Equator	Prime Meridian
distance from each other	same distance apart	Further apart near to the equator – the lines become closer together as they meet up at the poles.

25.

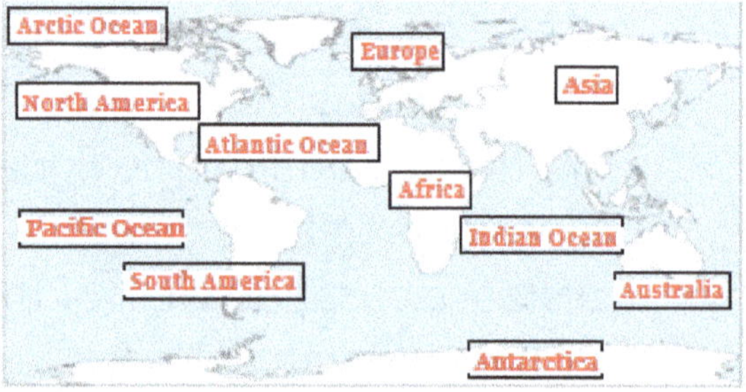

26. B. There are seven continents (North and South America, Asia, Europe, Africa, Australia, Antarctica) and five oceans (Atlantic, Pacific, Arctic, Indian and Southern/Antarctic)

27.A. The Mississippi River is located in North America, not Europe.

28.

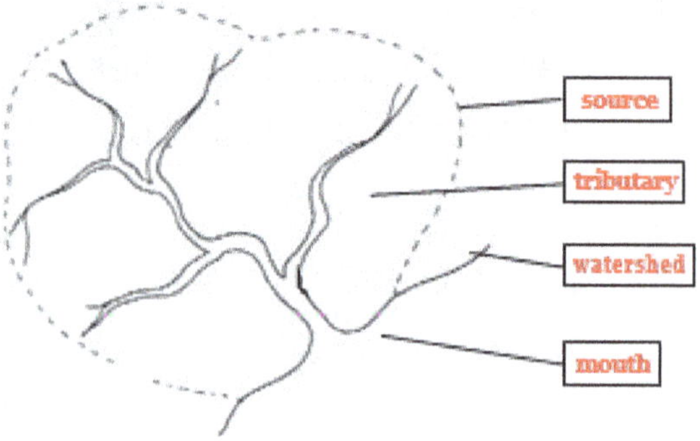

29. D. source, tributary, bank, mouth

30. C. Madeira is a tributary of the Amazon.

31. B. The global grid address would help issue a tsunami warning to the proper locations.

32. A. Using the global grid address by asking the captain for their latitude and longitude coordinates would allow rescue crews to locate the ship easily.

TERM 2: UNIT 3

Theme: Living Together

1. cItizen-B; leader-E, democracy-K; cabinet-I; government-J; parliament-H; opposition-C; senate-L; monarchy-G; constitution-F; vote-D; constituency-A;

2. **C.** government, constitution, democracy, monarch

3. Local government is a body of parish representatives elected by the citizens of each parish every two years. This governing body serves the citizens of the parish by providing services such as fire services, roads, markets, parks, recreational centres, street lights.

   Central government is the main government body in Jamaica. It is made up of three branches: the executive, legislature, judiciary.

   Richie should therefore go to his local government representative.

4.

Roles	Central Government	Local Government
They create policy on matters such as health, education, finance.	X	
They provide help for the poor, homeless, aged in communities.		X
They implement and change laws in the country's best interest.	X	
They maintain street lights.		X
They are responsible for beaches and public parks.		X

5. **C.** The people of Jamaica elect leaders in both the local or central government.

6. **D.** be in possession of a valid work permit

7. B and D

8. C

9.

Activity	Supports becoming a responsible citizen
Serve on a jury	√
Have multiply unpaid tickets for traffic violations	
Volunteer to help disabled children.	√
Form a group that would promote a law to force everyone in their communities to join the local community club.	

10. A

11. C

12. possible answers:

Rights	Responsibilities
right to life	reporting persons who commit murder
not being arrested without a cause	giving evidence to prove someone is innocent
right to have your own opinion on matters	do not share opinions that will cause public mischief.
assemble peacefully	protecting public assembly spaces
protection from discrimination	don't hate people because they are different from you
the right to free speech	respect the speech of others
the right to a fair trial	seek legal representation

Rights	Responsibilities
the right to vote	vote in council and general elections
the right to worship in any religion	ensure public worship does not violate the rights of others
the right to freedom of movement	show respect for curfew hours

13. A.

14. B.

15. Possible answers:

Name	How the position is acquired
Example: the governor-general	appointed by the monarch who he represents as head of state in ceremonies.
1. prime minister	appointed by the governor-general after the outcome of a general election
2. deputy prime minister	appointed by the governor-general on the advice of the prime minister
3. leader of the opposition	appointed by the governor-general after the results of a general election
4. members of the cabinet	appointed by the governor-general on the advice of the prime minister

15B. Answers may vary.

16.

Skills and Qualities
honest
effective communicator
visionary

17. Answers may vary.

18. Possible answers may vary

Ministry	Goods or Services
Culture, Gender, Entertainment, and Sport	builds new sports venues
Economic Growth and Job Creation	provides tax incentives for a company to make new jobs
Health	provides healthcare at hospitals and clinics
Justice	ensures the constabulary force is funded to provide protection for citizens
National Security	provides for armed forces
Science, Energy, and Technology	works to study the impact of climate change on the island
Transport and Mining	works to regulate mining and to ensure it is conducted sustainably

19. A, B, D

20. Answers may vary.

21. Answers may vary.

22a. Possible answers:

- They have identified the problem—need for a house.
- They have proposed a solution—find a cheaper house.
- They have dialogue with everyone involved.
- They will take everyone's feedback into consideration, then make a final decision.

22b. Answers may vary.

22c. Answers will vary.

23. C. Instituting laws and policies to ensure that all constabulary force members are held accountable for violating the rights of the citizens.

24. C

40-QUESTIONS' TEST ANSWERS AND EXPLANATIONS

'Success is no accident. It is hard work, perseverance, learning, studying, sacrifice and most of all, love of what you are doing or learning to do.'

— Pelé, Brazillian pro footballer

40-QUESTIONS TEST #1

1. **A.** Amerindians
2. **A** and **C.**
3. **D.** Indians, Diwali
4. **D.** They were free to leave the plantation at any time.
5. **D.** Jamaica, Trinidad and Tobago
6. **C.** location
7. **C.** burn it privately
8. **B.** protection
9. **A.** all products made in Jamaica.
10. **C.** order from emancipation to independence
11. **C.** founder of UNIA
12. **D.**

Symbol/Emblem	Description
	The coat of arms show a male and female Taino Indians standing on either side of the shield which has a red cross with five golden pineapples on top of it. The crest is a Jamaican crocodile which sits above the Royal Helmet and mantlings. It bears the motto "Out of Many, One People".

13. **C.** ethnic groups, immigrated, colonisation, emancipation, African
14. **B**. X – plain, Y – mountain, Z – hill
15. **C.** has the lowest average temperature
16. **C.** Blue Mountain is the highest mountain peak in Jamaica. The higher the altitude, the lower the temperature.
17. **D.** ranked based on height
18. **A.** I. true II. false III. true IV. false
19. **B.**

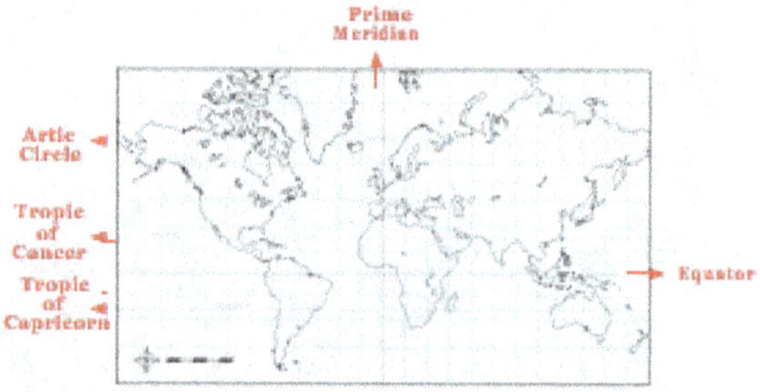

20. **C.**
21. **A.**
22. **B.** ship records between 1845 and 1916
23. **B.** i, ii, v, vi and vii
24. **A.** soil erosion

D. loss of habitats for animals

25. B. their cultural practices

C. business prowess

26. C. universal adult suffrage

27. A. own a Jamaican passport

B. vote in a general election

28. B. house of representatives

29. B. The ex-enslaved saw them as threats.

C. The Chinese were paid less than the ex-enslaved.

30. C. rotation and revolution

31. B. Aruba

32. D. Overseas region.

Aruba is located in the Caribbean and is one of four countries which constitute the Kingdom of the Netherlands centered on the continent of Europe.

33. C. is taken and governed by another country

34. A. Jamaica and Barbados

35. A. St. Lucia and Grenada

36. A. executive branch

37. C. judicial branch

38. A. related to the sea or ocean

39. D. increased obesity

40. A. recycling

"Recognize yourself in he and she who are not like you and me."
— Carlos Fuentes
What does this mean? Come wid mi!

40–QUESTIONS TEST #2

1. **D.** Africans
2. **A.** colonisation
3. **D.** East Indian, Sari
4. **C.** Chinese
5. **C.** Toussaint L'ouverture

   D. Jean-Jacques Dessalines
6. **A.** self-government
7. **D.** playing it every morning and evening
8. **B.** indentureship
9. **A.** emigrant
10. **A, B and D.**

    Push factors:

    poverty, lack of job opportunities, low wages, poor health care, famine, natural disasters, overpopulation

    Pull factors:

    better job opportunities, higher wages, free passage back to his country, shelter, better medical care
11. **D.** British and Indians
12. **A.**

Celebration/ Festival	Religious	Secular
Divali	✓	
Crop Over		✓
Hosay	✓	
Chinese New Year		✓
Eid ul Fitr	✓	

13. **B.**

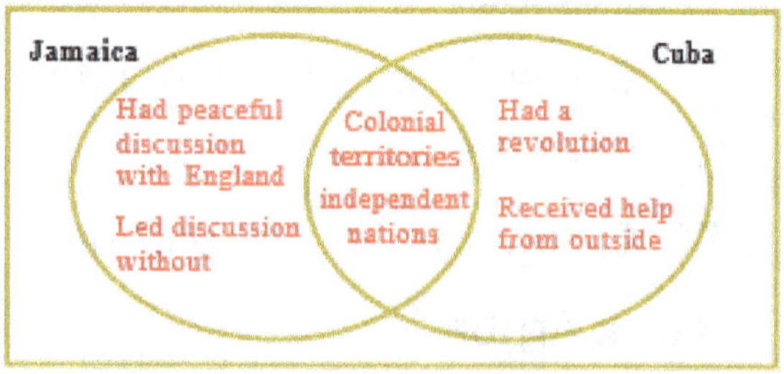

14. **A.** Cuba, 1902; Haiti, 1804; Jamaica, 1962
15. **B.**

Country	Independence status	Year of independence
Anguilla	dependent	n/a
The Bahamas	independent	1973
Barbados	independent	1966
Montserrat	dependent	n/a
Jamaica	independent	1962
Cuba	independent	1902
Cayman Islands	dependent	n/a

16. B. Norman Manley
17. **C.**

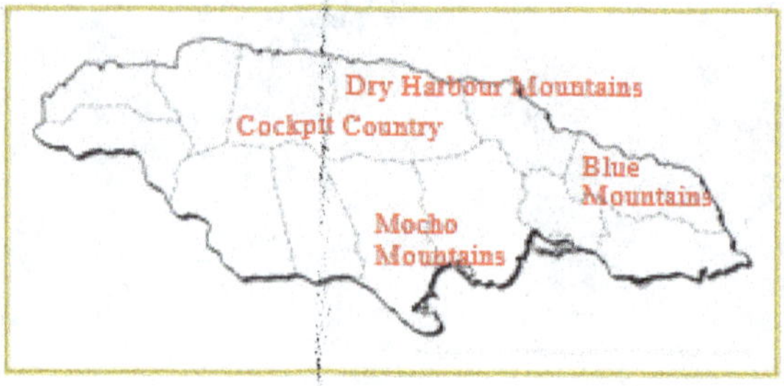

18. C.

Human activity	Environmental Impact
1. construction of houses (settlements)	landslides/soil erosion
2. cultivation of crops (farming)	soil erosion/landslides
3. cutting down trees (deforestation)	loss of habitats for animals

19. D. soil erosion

20. B. lines of latitude are all parallel

21. D.

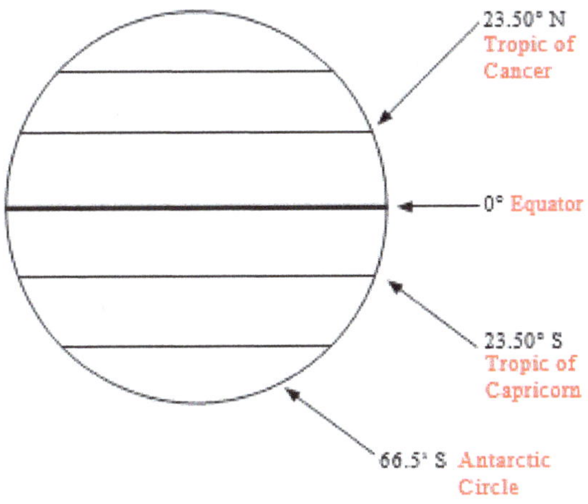

22. A. Equator

23. C. Africans

24. B. To show respect for the national emblem.

25. C. It is very durable.

26. C. There is no river in Antarctica.

27. A. There is a river in Antarctica called the Onyx River

28. A. Yangtze in Asia

B. Mississippi in North America

29. B. Consumer Affairs Commission

30. B. ii and vi

31. A. dual citizenship

32. A. citizens

33. D. Chinese New Year

34. B. II and IV

35. C. protecting your neighbour's property

36. C. The Court of Appeal

37. A. every community has a mayor

38. D. Right to privacy

39. health care, water, energy, security, education

40. C. The Bahamas

40–QUESTIONS TEST #3

Multiple Choice:

1. **C**
2. **A**

Adult male citizens after 1962 had the most civil rights. Slaves had no rights, while indentured servants had some rights. Personal servants had more rights than indentured servants but fewer than citizens, especially adult men.

3. **A**

Answer choice A is the best interpretation of the Jamaican national motto. Answer choice B is true but not the meaning of the motto and C and D are false statements.

4. **D**

Christopher Columbus was the Italian explorer who first came to Jamaica. King George and Prince Albert were British, and Bob Marley was a musician.

5. **D**

The geography, economy and history of a place are equally important to include in an informative report.

6. **B**

The Sam Sharpe revolt, which helped pave the path to emancipation, took place in 1831. Apprenticeship was implemented in 1838, not the right to vote. In 1944, universal adult suffrage was granted to all adult citizens. In 1962, Jamaica gained its independence from Britain, not Spain.

7. **A**

Many believers refer to Diwali as the Festival of Lights because it celebrates light and goodness. New Year's Day celebrates the beginning of a new year and Crop Over celebrates the end of the harvest.

8. **C**

Information contained in encyclopedias and textbooks may not be current. In addition, these references only provide cold, hard facts. However, a personal diary would reveal his thoughts and feelings and his perspective on the events and people in his life. A candid (not posed) picture could provide clues about his life and insight into his character. Still, the personal diary is the best reference, second only to a personal interview.

9. D

Emancipation, which ended slavery and freed the enslaved, had to happen before universal adult suffrage, which gave all adults the right to vote. A person must have freedom in order to have rights.

10. D

Answer choice D is the best answer because it is an easily proven, commonly known fact.

11. C

Higher elevations have cooler temperatures and more rain than lower elevations; therefore, neither answer is true.

12. D

An isthmus is a landmass with water on two sides and connects two larger landmasses. A gulf is not a landmass. An island is surrounded by water and a peninsula has water on three sides.

13. D

There are several ways other than by birth for someone to become a Jamaican citizen. A is false. No one has complete freedom to do whatever they choose. One person's rights ends where another person's rights begin. B is false.

14. B

The passage says longer, not shorter, periods of drought are part of climate change.

15. B

Employment and income are privileges or opportunities, but they are not rights. Jamaican citizens have the right to privacy and protection from other people and/ or governments.

16. C

The English arrived in 1655 and attacked the Spanish. They won the battles and then they let the enslaved go. They stayed in Jamaica for years by establishing settlements.

17. B

Besides Jamaica, the Dominican Republic is the only country listed that is independent. This means that this nation governs itself. The other countries are dependent. They are ruled by other countries.

18. C

The pineapple fruit was transported to Jamaica from South America by the island's first inhabitants, the Taino Indians. That is why it is included on the shield.

19. A and C

The national flag should not hit the floor. It is supposed to be flown on the flagpole. When being stored, the flag should be folded in a particular way and then put away safely.

20. C

'President' refers to one person who works for the government, while the others are the branches of the government, which consists of many people.

21. D

Bays are small bodies of water that are partially enclosed by the land. They eventually empty into the larger body of water, which is the ocean.

22. B. 2.4 to 1

Ratio of Ocean: Continent
361 sq km : 150 sq km
2.4 : 1

(each side was divided by 150)

So the oceans covers an area of 2.4 times the area covered by the continents.

23. D. Used to locate places

24. D

The key fact known about Norman Manley's contribution to Jamaica's independence is that he was elected as the chief minister. This allowed him to help negotiate with the British government, the ruling country, for Jamaica's independence.

25. C

The Blue Mountains were formed by volcanic activity. This means the soil will be rich in nutrients and minerals and is therefore fertile for growing crops.

26. A, B and D

In answer A, 'best' and 'all' are opinion words because they cannot be objectively measured. In answer B, 'interesting' is an immeasurable opinion. Popular, in answer D, cannot be measured. Answer C is the only statement that can be measured and proven or disproven; therefore, it is NOT an opinion.

27. B. Culture includes traditions and beliefs.

28. A and C

29. A and D

The three main branches of government are judicial, legislative and executive. Ancestry and population may influence the government but are not parts of the government.

30. A, C and D

Compromise is achieved when each person gives up something he/she wants and each person gets something he/she wants. Honesty, open communication and mutual respect are essential to great team-building. Gossip, however, will destroy a team and it is not usually honest.

31. A and C

The most common way to become a Jamaican citizen is by birth, either being born in Jamaica or being born somewhere else to at least one Jamaican parent. Answers A and C are correct. Answers B and D are false.

32. A and D

Blue Mountain Peak and John Crow Mountains are popular mountains in Jamaica. Mount Vesuvius is the volcano that wiped out Pompeii and Mount Everest, the world's tallest mountain, is located between China and Nepal.

33. B and C

Snorkelling and rafting can both be enjoyed in a tropical climate, while snow and ice would not last long in a tropical climate, so answers B and D are correct.

34. B and D

Answer: The source is the beginning of the river and the mouth is the end.

35. A and B

Atmospheric pollution can occur anywhere and affect anyone. So, answers A and B are myths or misconceptions. Because their immune systems are often inadequate and they frequently have respiratory problems, the very young and very old are more susceptible than most people. There are no boundaries or borders of atmospheric pollution, so everyone should try to help stop it.

36. A. 'Option A lists a set of coastal features. Coastal features are special landforms created where our landmasses meet our seas and oceans.

37. A. windward side - side of the mountain usually exposed to the wind

38. C. decision made through discussion and negotiation as a group

39. B.

40. A. on the coat of arms

How are you doing? During and after your PEP exams, always remember the six **P**s:

Proper

Planning and **P**reparation

Prevent

Poor

Performance.

Whatever your current scores, you still have time to make improvements.

Thanks for taking the journey wid mi.

GLOSSARY

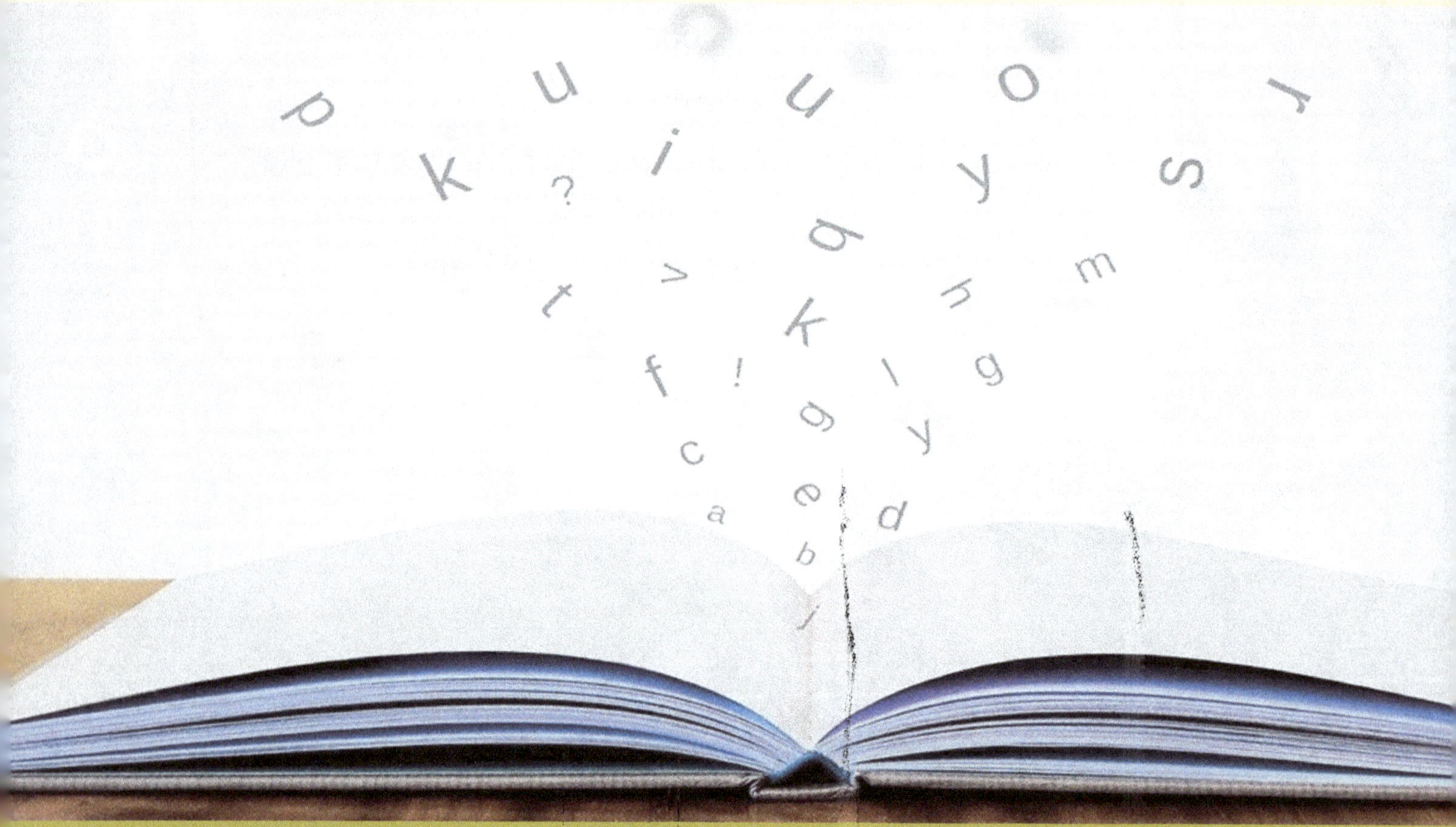

'One of the really bad things you can do to your writing is to dress up the vocabulary, working for long words because you're maybe a little bit ashamed of your short ones. This is like dressing up a household pet in evening clothes. The pet is embarrassed, and the person who committed this act of premeditated cuteness should be even more embarrassed.'

– Stephen King, *On Writing: A Memoir of the Craft*

A

adoption - the process of legally taking a child as your own

amendment - a change made to a law or agreement

anthem - a song with special meaning for a nation or organisation that is used on special occasions

archipelago - a series of islands close together

Attorney General - a lawyer appointed to advise the government on legal matters

B

bay - the smallest inlet of water protected by land

bearing - the central part of the coat of arms to which everything is attached. The bearing on the Jamaican coat of arms consists of the shield with a red cross and five golden pineapples.

bill - a document outlining a proposal by parliament

C

Cabinet - a small group of special representatives who assist the leader of the government closely

carnival - a festive event, usually held in the street, marked by entertainment, celebrations, processions, music and costumes

central government - the main governing body in the country

ceremony - formal celebration of a special event, usually following a structure and includes specific actions or words

citizen - someone who can legally live in a country through birth, marriage, adoption or other legal process and who has certain rights and responsibilities in that country

coat of arms - a design with a shield, crest and arms that have meaning to a nation or institution

colonial rule - government of a weaker nation by a stronger, wealthier and typically older one

commonwealth - an independent nation symbolically headed by a British monarch

constabulary force - a group of persons specially trained to keep people safe, make sure they obey laws and maintain order in a country

constituency - the people who are represented by an elected member of the government; an electoral district

constitution - a written set of rules or laws outlining how a country is governed, the powers and duties of members of government and the rights and responsibilities of citizens

consultation - a discussion where persons seek advice or share ideas before making a decision

continent - one of seven large landmasses on Earth, usually divided into different countries

contract - a legally binding agreement between parties that tells the duties and expectations of each party

crest - the highest point or something on top of the head of an animal; an image that represents a nation or institution

culture - the way of life, practices, traditions, festivals, beliefs (religious), language, creative arts etc. of a people

D

decision - a choice that is made after considering other choices

democracy - a government that allows the citizens equal right to vote for their representative

E

emblem - an object used to represent something of significance

emigrant - someone who leaves his or her country of origin to live in another country

ethnic group - a community or group of people who have common cultural practices and are from the same descendants

executive (arm of the Jamaican government) - the policy-making body of the government

F

festival - a special day or period, usually in memory of a religious event with its own social activities, food or ceremonies

flag - a fabric shaped and designed to symbolise a country, an idea or belief system (e.g. Rastafarianism)

forest reserve - large area of trees, other plants and wildlife that has been protected by the government and where activities that would destroy the area are not allowed

franchise - an individual or a group that is authorised to be an agent of a company's products or services

G

goods - physical objects that be bought or sold, used or consumed. They are also called products or commodities

government - the body that sets rules and laws, makes decisions and policies for a country

Governor General - person appointed to represent the king or queen as head of state

gulf - an inlet of water that is almost surrounded by the sea

H

heritage - cultural things such as art, religion, art forms, traditions, practices and activities passed down through generations

hill - a landform that rises slightly higher than the area around it and often has a round top.

humanely - treat with kindness, compassion and respect for life

I

immigrant - a person who leaves his/her country and lives in a foreign country

indentured servant - an immigrant labourer who works for a fixed time in exchange for wages and basic provisions such as housing and land or a return trip home at the end of the contract

indentureship - an economic system in which workers from other countries are contracted to work for a period

independence - a country's freedom to govern or rule itself without interference from another country

influence - have an effect or be able to change the outcome of events or others' behaviour

integration - countries coming together to share knowledge and resources so each country benefits

integrity - the quality of having good morals, honesty and trustworthiness

island - land surrounded by water

isthmus - a narrow piece of land that connects two larger pieces of land

L

lake - a body of water that is completely surrounded by land

landforms - the natural geographic features of an area

leader - someone with influence or authority over others; one in charge who makes decisions on behalf of others

legal - according to the law

legislator - someone with the power and ability to create laws

local government - a body of officials elected to serve the people at the parish level

M

migration - movement from one location to another area to settle

minister - an elected representative who heads different government divisions called ministries

monarchy - a country ruled by a king or queen

motto - a statement that summarises the most important beliefs or values of an institution, group, individual or country

mountain - a landform with high peaks and steep or rugged slopes

mountain range - a string of mountains that are close together

N

nation - a country with its government and land; people from the same country

nationhood - the state of having independence and national identity

naturalisation - the legal process of making someone from another country a citizen

O

ocean - the largest type of water body on Earth

opposition - people in parliament who are elected by constituents but are not part of the team that makes up the ruling party.

P

parliament - a group of law-makers that represent the people and oversee the government

patriotism - strong support and love for one's country

peninsula - a piece of land surrounded by water on three sides

plain - a flat, low-lying area of land often used for farming

plateau - a high, flat area of land

political party - an organised group of people with shared beliefs and ideas about politics and who present candidates to be elected to form part of the government

Prime Minister - an elected person who is the leader of the country

protocol - the correct or appropriate procedures to follow, especially on official occasions

pull factor - a benefit or motivator that draws a person to a place

push factor - an unpleasant condition or circumstance that encourages an individual to migrate

R

region: a large or continuous area with common physical and human features that are different from neighbouring areas. Physical features include mountains, valleys and climate while human features include population and culture

relationship - the way which two or more people connect, relate with or are involved with each other

responsibility - duty or role someone should carry out usually because it is part of a job or is expected

revolution - a violent overthrow of a government in favour of a new one

right - a benefit or privilege to which someone is morally or legally entitled.

river - a running channel of water that flows to the sea

S

sea - a large body of salt water that is smaller than an ocean

self-government - freedom to make decisions for and manage your own country; freedom from outside control

senate - a legislative body that has a little more power than the house of representatives or lower body

slope - the side of a mountain or hill

source - material, such as a book or website or person that provides useful information for studying or doing research

summit - the top of a mountain

symbol - anything that is used to represent something else; something physical or tangible that is used it to stand for an idea, quality or some other abstract concept

T

taxes - money that must be paid to the government to cover the country's expenses

trade - the action of transferring services and goods between/among persons, companies or countries, usually in exchange for money or another product

trade union - an organised association that negotiates for workers and protects their rights universal adult suffrage - the right of all adult citizens to vote without discrimination

V

valley - low-lying area shaped like a U or V, formed between the sides of mountains

vote - to cast a ballot to choose a person running for office

NOTES

www.ingramcontent.com/pod-product-compliance
Lightning Source LLC
Chambersburg PA
CBHW081805200326
41597CB00023B/4148
9781732836952